Breakthrough to Math

LEVEL 3

TEACHER'S GUIDE

by Stephanie Irwin

Breakthrough to Math Level 3
Teacher's Guide
ISBN 978-1-56420-992-4

New Readers Press
ProLiteracy's Publishing Division
101 Wyoming Street, Syracuse, New York 13204
www.newreaderspress.com

Printed in the United States of America
24

Proceeds from the sale of New Readers Press materials support professional development, training, and technical assistance programs of ProLiteracy that benefit local literacy programs in the U.S. and around the globe.

Contributing Writer: Practical Strategies, Inc.
Developmental Editors: Beth Oddy, Terrie Lipke
Design and Production Director: James P. Wallace
Production Specialist: Maryellen Casey
Senior Designer: Carolyn Wallace

Table of Contents

Introduction

The Adult Learner

The *Breakthrough to Math* series was designed to accomplish two purposes: to meet the special educational needs of adult learners and to help meet some of the particular needs of adult education math teachers.

On the whole, adults returning to the classroom environment share certain characteristics, problems, needs, etc., which frequently include:

Negative Self-Image With Regard to Educational Skills and/or Learning Abilities Most adults, to one degree or another, feel that they failed insofar as completing educational programs via "normal" channels is concerned. They may, in fact, be aware of external factors that might have contributed to their inability to finish school as per traditional modes. However, most adults reenter the classroom environment very conscious of and self-conscious about their past educational failure. They hold themselves ultimately responsible for that failure. Adult students, then, frequently return to the classroom anticipating continued failure.

Personal Responsibilities and/or Schedule Conflicts Many adults returning to the classroom environment experience transitioning difficulties. As adults, they have non-school responsibilities, e.g., job, family, etc. Not infrequently, this reordering or transitioning process poses problems or conflicts for the returning adult students which may include or be evidenced by: irregular class attendance; in-class fatigue resulting in poor concentration, short attention span, etc.; incomplete homework assignments; and progressive in-class frustration leading to dropout.

Math Anxiety Of all subjects/skills development areas encountered by returning adult students, math is the area most frequently cited by the adults themselves as likely to cause them the most difficulty. Many adult students have had, at best, frustrating experiences with math subject areas in their public school experiences. Further, most are convinced that their negative experiences were primarily due to their own personal lack of ability. In actuality, adult education program data substantiates that the majority of adults enrolled in math instruction programs do not experience insurmountable difficulty with math skills development. What is required, however, is that the adult education math instruction, materials, and teaching techniques be tailored to the needs of adults. This is to ensure that the adult is not again subjected to the kinds of negative classroom situations that he or she experienced in previous educational settings. Given appropriate materials and instructional approaches, adults begin to realize that math is, in reality, a series of ordered and understandable skills that they, too, can master.

In-Class Stress Caused by Peer Competition All too often, the returning adult student has experienced frustration in the public school environment, due to an inability to keep up with his or her peers in math classes. Frequently, the adult's emphasis and energy have been directed toward at least seeming to keep pace with others rather than toward actual development of math skills and comprehension of math concepts. Again, the success of an adult learner in an adult education math class will be proportionate to the degree to which instructional materials and teaching techniques stress skills accomplishment according to individual student learning pace, style, etc.

Teacher Dependence/Lack of Awareness of Personal Role in and Responsibilities for Learning Adults returning to the classroom environment have generally not had previous educational experiences that have helped them realize the extent to which they can contribute to and assume responsibility for their own learning. Too frequently, they enter adult education classes anticipating that all knowledge will come from the teacher. Adult education math classes may provide the setting and the opportunity for adults to learn to assume responsibility for their own learning and to develop a sense of educational independence which may carry over to other settings. This occurs, however, only if the instructional materials and techniques allow for and encourage the adult's active—rather than passive—participation in the learning process.

Difference Among Adult Students at Program Entry With Regard to Skill Levels, Learning Rates, and Styles Adults learn in different ways, at different rates, and at different times. Further, adult students reenter the classroom environment with varying skills and varying skill development needs. It is therefore most essential that materials and teaching techniques allow for and support, where possible, the learning pace and style of individual students and that the classroom experience permit students to realize and build upon existing skills while promoting the development of new skills.

Adult Educator's Concerns

As a rule, the needs, concerns, and problems of adult education instructors differ from those of K–12 and/or higher education teachers. In some cases, these differences may simply be a matter of degree; in others, these differences are unique to the adult education experience. Differences may include:

Heterogeneous Class Groupings Frequently, adult education teachers face classes that include adults ranging in age from 16 to 60-plus years. Further, student skill levels at program entry often differ drastically. The teacher's ability to provide individualized instruction for these diverse groups will be directly affected by the kinds of instructional materials available.

Staff Resources / Turnover / Available Classroom Times Frequently, adult education instructors teach in programs that operate on part-time schedules, have significant staff turnover rates, provide ongoing or open enrollment for students, and attempt to provide individualized programming for adult students.

Therefore, adult education programs need instructional materials that

- can be easily digested by new instructional staff,
- lend themselves to individual student skills development needs/attendance schedules/learning rates and styles,
- provide easiest access to skills development for students and teachers with limited available classroom time,
- provide a consistent and sequential approach to subject matter for both students and staff, and
- promote the effective and efficient instruction of numbers of students, on an open enrollment basis, by limited teaching personnel.

How to Use the Series

The point of program entry is critical for returning adult students and their adult education math instructors. It is at this point that the student feels most insecure about his or her decision to return to school and chances of success in the program. It is at this point that the teacher must not only determine the most appropriate starting level for the student but also simultaneously avoid causing additional stress and anxiety in the student. It is also at this point that the teacher will lay the base for the important teacher-student relationship which will endure throughout the student's program involvement. Most frequently, teachers are working with numbers of students at this time and so have a need for materials that are easy to utilize and are self-pacing.

The Locator and Placement Inventory

The Locator and Placement Inventory tests are most appropriate for both teacher and students for several reasons. The tests provide students with a sense of commonality—all students begin instruction with the same kind of instructional materials. The materials are self-paced and student-directed to the degree preferred by the teacher. As students begin to indicate the computations they cannot perform on their own, the teacher begins to interact with individual students as needed. The teacher begins to provide positive reinforcement with regard to those skills the student has mastered and begins to nonthreateningly indicate how these skills can be built up in a gradual manner and at a pace that is most suitable to the student's individual learning style. Both student and teacher participate in reducing the student's math anxiety while simultaneously formulating a positive and supportive student-teacher relationship.

The Locator and Placement Inventory tests allow students to quickly advance to the appropriate point of needed skill development. Feedback from these tests results in students moving forward to the particular skill books that will address individual skill needs. This will promote an important sense of self-confidence in the student at the outset of program involvement.

These tests provide the teacher and student with an organized method for keeping track of the student's progress in skills development. It is particularly important that the adult have available some method of tracking his or her own progress through the program. Knowing what has been accomplished and what is yet to be accomplished helps the adult maintain independence and a sense of responsibility for, and participation in, his or her own learning.

The Skill Books

All of the skill books in the *Breakthrough to Math* system have a similar appearance and follow the same lesson style. This is an important factor with regard to providing students, particularly those working at lower levels, with a sense of doing the same kinds of work as other students. While this may seem an inconsequential point on the surface, it is nonetheless often a positive and supportive factor as far as students' attitudes are concerned. Lower level math students are often painfully aware of and embarrassed by the fact that their materials look easy or immature compared to the more sophisticated appearance of materials utilized by more advanced students.

The pre-test and post-tests included in the books provide teachers with a readily accessible means of doing ongoing assessment of skills development needs and skills attainment. Practice exercises and accompanying answer keys found in each book contribute not only to an organized, sequential, and accessible learning experience for both student and teacher, but also to the student's development of a sense of responsibility for his or her own learning. Proper use of the answer keys affords the student an opportunity to keep in touch with his or her thinking processes while participating in skills development activities. Correct answers reinforce a sense of accomplishment and faith in the student's ability to read, comprehend, and apply skills information. Incorrect answers will cue the student as to when interaction with a teacher is needed to clarify and/or reinforce problem areas. The books, then, encourage the student's active and responsible participation in his or her own learning processes. They also free the teacher to work with individual students as needed. Students progress as quickly as they are able to or as slowly as they need to, so the teacher's time is put to the most effective and efficient use.

It should be noted that the *Breakthrough to Math* series is not intended to be the sole source of skills development for students. The system should be considered open-ended. Teachers are encouraged to plug into the system those supplemental materials and activities that are part of the existing program curriculum, whenever and wherever appropriate. For some students, *Breakthrough to Math* will provide enough practice exercises for needed skills development. For other students, the workbook will provide additional activities to reinforce and practice skills.

The *Mastery Checkups* and Post-Tests

As with the individual skill books that comprise *Breakthrough to Math,* the *Mastery Checkups* and post-tests are not expected to meet all possible needs of all possible adult education math students. While they will meet the needs of some, other students may require supplemental tests and/or skills development checkup activities.

The *Student Profile*

The obvious value of the *Student Profile* is that it will provide teachers with an organized method for tracking individual student progress in skill mastery. However, it should also be noted that it will provide the same for the student. Existing adult education program data supports the indication that adult students tend to respond positively when they have an understanding of where they are in the math curriculum. Knowing what they've accomplished as well as the work that remains to be done promotes a sense of continuity. It helps adults plan for effective use of in-class and homework time. It is also a key factor in helping them realize the extent to which they may actively participate in, assume responsibility for, and exercise some positive control over their own learning experiences.

Barbara Banks
Series Developer

Math Phobia

Mathematics can be very frightening to people who have not mastered it. The fear often stems from childhood experiences. The biggest difference between those who like math and those who fear it is success. Students who cannot do math often believe they are stupid and have low self-esteem.

Throughout your work, your student's ego must be a major focus. It is important not to forget the student while focusing on the task.

Many teachers are not comfortable teaching math. Even those trained in the subject may feel insecure teaching certain topics. Teachers may also be frustrated by negative behaviors of students who are slower to grasp concepts and processes than teachers expect them to be. It is important to remember that ideas which seem to be common sense or simple to you may cause difficulties for students with math phobia.

To avoid these feelings

- thoroughly prepare each lesson ahead of time,
- admit to students that this may be difficult material,
- use this *Teacher's Guide* to prepare and present alternative teaching approaches, and
- model problems for all exercises before students start working them.

Answering the student's questions and patiently showing the student a process more slowly, several times, line by line may help you and the student successfully get through a difficult skill and gain confidence. Repeating work on the same problem until the student has mastered it and feels comfortable will help both of you approach the next problem with a greater chance of success. Model carefully all work for the student; surprises are not productive for a student with math phobia.

How Students Learn

Students learn in different ways. Some students are visual learners. They see and remember as if they're looking at a chalkboard in their minds. For these students, seeing the concept and/or writing and looking at the problem may be the easiest way for them to learn. Other students are auditory learners. They hear and repeat a fact to memorize it. You may be able to tell which way an individual student naturally receives and retrieves information by watching and listening carefully as he or she attempts to remember something. Place six different items on a tray. Ask the student to remember the items. Cover the tray and ask the student to list the items. Note if the student squints, stares off into space, or closes his or her eyes. If this is the case, that student learns best in a visual mode. An auditory receiver may look at the covered tray and may think alphabetically or rhythmically when trying to remember the items. You may notice the student mouthing words. This indicates that this student learns best using an auditory mode. Visual students receive better by watching you do a problem and then trying a problem on paper themselves. Auditory learners receive better if you discuss the problem together, speaking about the important or tricky aspects of it.

Go Slowly Many things in math (facts, formulas, processes) must be memorized. Small portions of related items are most easily committed to memory. Adding items just as slowly is vital.

Overloading a student with too much information at any one time may lead to feelings of frustration. Repeating an earlier success that has been slowly and carefully developed convinces students that they can successfully learn new material.

Be Open Encourage students to ask questions or to ask you to slow down if they feel you are moving too fast. As long as students have unanswered questions or cannot keep up, effective learning will not take place. If you do not know an answer, admit it. Assure the student that it was a good question and that you will find the answer.

Stress Correct Process Correct answers will result from following the correct processes. Mistakes or errors should be corrected together—with the student. It is extremely important that you stress the positive (especially correct process) rather than the negative when working with a student who has math phobia. Discuss the number of problems the student did correctly, and how, before going over those missed. Do not use a red pencil for correcting exercises.

Reading and Writing Remember that your students may have difficulties with reading and writing that contribute to their difficulty with math. Be sensitive to this. Go over directions and problems together whenever necessary.

Using the Diagnostic Tests

The *Teacher's Directory* contains an overview of the *Breakthrough to Math* series as well as the Locator test and the answers to the Placement Inventory and Mastery Checkups that you will be using with this series.

To place your students appropriately in the series, administer the Locator test. This test, found in the *Teacher's Directory,* contains questions that are intended to quickly locate that portion of the Placement Inventory test where each student should begin.

Quickly and casually glance at the Locator when the student has completed it. Without comment, hand each student the appropriate Placement Inventory test after you have circled the beginning question as listed under "Locator Referral" in the *Teacher's Directory.*

The questions missed, if any, in the Placement Inventory will indicate to you which level and which book within that level are appropriate for the pre-test for that student. The pre-test of that book will show you the chapter where your student should begin. Students who answer correctly all the problems on the pre-test should take the pre-test in the next book.

Scaffolded Learning *Breakthrough to Math* has been designed with a very systematic approach to skill building. For example, *Mastery Checkup 3* is used after completion of Level 3. The chart below shows the mathematical skills that are taught in Level 3 of the series. As the chart shows, mathematics is a skill-building or scaffolded-learning process. Each new concept builds on a previously taught concept.

When a student has difficulty with a task, look for any preceding skills that the student may not know. The chart should help you understand the interdependence of these skills, pinpoint weak areas, and begin to develop the math skills.

How to Use the *Teacher's Guide* and the *Workbook*

The *Teacher's Guide* and the *Workbook* for Level 3 are designed to assist you by offering a variety of methods for teaching a concept; by providing background skill-building materials; and by providing additional practice for students who need it.

The *Teacher's Guide* has been developed to parallel the skill books chapter by chapter and page by page. For example, if you have been referred by the pre-test to Chapter 2 in Skill Book 1, turn to the Table of Contents of this *Teacher's Guide* for the appropriate page that will give you teaching suggestions, alternative teaching approaches, and extra practice suggestions.

The answers to the worksheets in the accompanying *Workbook* are located in the back of the book so that students may check their own work. It is important for students to check each problem as it is completed so that they do not continue to practice an incorrect method. It also helps develop students' confidence.

Some General Instructional Strategies

The following strategies may prove helpful in getting your students to see past any math anxiety they may have so they can begin to think about how math works. The Think-Aloud technique especially can demystify math and solving problems.

Typically, a teacher approaches a problem as though he or she already knows what to do. The student thinks that the process for solving the problem must

Level 3 Math Skills

Signed Numbers	Solving Equations	Word Problems	Exponents, Roots, and Polynomials	Algebraic Graphs
Positive and negative numbers Adding Subtracting Multiplying Dividing Using more than one operation	Translating expressions to words Writing expressions Solving equations Using inverse operations Combining like variables Working with literal equations	Using formulas Writing your own equations Ratios Proportions	Factors Exponents Square roots Like terms Processing monomials Processing polynomials	Locating coordinates Naming coordinates Plotting equations Solving two equations

be obvious because the teacher knows it. The student then feels stupid or frustrated, or most often, both. The first step that students should learn is to ask, "What is this problem asking me to do?" It short-circuits their frustration by giving them a concrete process to begin to answer the question. The graphic organizer introduced in this section provides a framework for asking and answering this question.

Think-Aloud Technique

The Think-Aloud technique is an excellent way for teachers to help students improve their critical thinking and problem-solving skills. Typically, when students see solutions to problems in textbooks or demonstrated by teachers, what they see is a step-by-step application of math operations that leads directly to the correct solution. This can be frustrating to students because it appears that the process for solving the problem should be obvious and their own abilities must fall short because it wasn't obvious to them.

What students miss is the thinking process that goes into reading the problem, analyzing what the problem is asking, identifying important information, thinking through the steps and the operations needed, and solving the problem. The process includes making and identifying mistakes. In the Think-Aloud technique, the teacher models critical thinking and problem-solving skills by periodically "thinking aloud" as she works the problem. Students can then observe how someone interacts with a math problem to solve it successfully.

Before Using the Think-Aloud technique

Step 1: **Select a problem that is challenging for students but that involves math operations they are already familiar with.** You want students to concentrate on the critical thinking and problem-solving strategies you use to solve the problem without being distracted by new operations.

Step 2: **Review the problem and select the strategies you will use.** Analyze the problem to identify what students need to be able to do to solve it, and where they are likely to encounter difficulties. Identify three to five strategies to use in solving the problem that directly impact students' areas of weakness. Rehearse how you will use the strategies and what you'll say. Preparation will help you focus on the specific strategies and present them clearly.

Using the Think-Aloud Technique

Step 3: **Clarify the problem by putting it in your own words.** This is the first step a good mathematician takes, but many students don't do this. Restate the problem in your own words, clearly identifying what you are being asked to find.

Step 4: **Solve the problem and model the strategies you identified.** Model the strategies by stopping periodically and thinking aloud about the problem. It's important to rehearse when you'll use the strategies and what you'll say, rather than demonstrating this on the fly. Preparation will help you focus on the specific strategies and present them clearly. It is all right and even preferred to demonstrate the thinking process as not perfect. You can try strategies that are wrong, for example. Just explain why you tried them and how you realized the strategies weren't correct. While you're using the Think-Aloud technique, have students follow along and write down the different strategies you try.

Reviewing and Extending the Think-Aloud Technique

Step 5: **Identify the strategies and discuss how they were used.** For short problems, do this step after you've finished solving the problem. If it's a longer, more complicated problem, you can point out the strategies when you use them, and then discuss them in more detail when you have finished solving the problem. Have students identify why a specific strategy was useful. Ask students if they can think of other problems they have seen where that strategy would be useful. Ask them to think of other strategies that you might have used.

Step 6: **Give the students problems and have them do the Think-Aloud technique using the same strategies.** You can do this in two ways. You can solve problems as a class, calling on different students to think aloud to the entire class. You can have students work in pairs and think aloud to each other while you go around the room and listen.

NOTE: Having the students use the Think-Aloud process either in explaining answers in a group or in pairs allows you to identify students' strengths and weaknesses and what you need to teach or reteach.

The Problem-Solving Process

Math is full of facts and operations that students have to remember and understand. When you consider that these facts and operations can then be contextualized in an endless array of word problems, it's no wonder that word problems are intimidating. There will always be word problems with contexts and wording that are unfamiliar to the student. What students need is a consistent process they can apply to solving any word problem they encounter.

In *Breakthrough to Math,* students learn to always ask questions when solving a word problem.

Students learn this problem-solving process in *Word Problems With Whole Numbers*. Introducing the process early on helps students become familiar with it and confident in using it. This will alleviate some of their anxiety as they encounter new word problems throughout this level.

Problem-Solving Graphic Organizer The graphic organizer on the next page will help students apply the problem-solving process to algebra word problems. This organizer asks six questions which help students to understand and solve word problems in Book 3. You'll see on the left side of the graphic organizer are the six questions (steps) in the process. On the right are sentence starters to help students respond to the questions. Requiring students to write their responses to the questions enhances their understanding of the process and helps you see their thinking. You can make copies of the graphic organizer for students to use when solving word problems.

Writing Word Problems The purpose of teaching students to write their own word problems is for the students to

- become familiar with the language of math,
- recognize where math occurs in everyday life, and
- practice problem-solving skills.

The goal is to help students see that word problems are not made up out of thin air. They have a context in real life.

Steps to Writing Word Problems:

1. Have students work in pairs.
2. Ask each student to make up a simple math problem and write it on a piece of paper, without the answer. Example: $10 \div 2 =$
3. Ask students to exchange math problems with a partner.
4. Ask students to make up a context for their partner's math problem. For the example above, the context could be $10 divided between two people.
5. Once they have a context, ask students to write a word problem to go with the math problem. Example: Mrs. Johnson has two children. While at the mall, she wants to give each of them money so they can shop on their own. She looks in her purse and sees that she has $10. How much money will each child receive?
6. Ask each student pair to exchange their two word problems with another student pair.
7. Have students work in pairs to solve the new word problems.

Writing Algebraic Equations for Real Life The purpose for teaching students to write algebraic equations for real life is for students to practice writing and solving algebraic equations and for students to connect algebraic expressions to concrete examples. The goal is to give students a sense of the mechanics of math: equation writers create problems that reflect the real world.

How to Write Algebraic Equations for Real Life

1. Explain to students that they are going to answer the problem: How many plastic cups can be stacked in a file cabinet drawer C inches tall? (Measure the cabinet ahead of time and use that value.)
2. Display several stacks of two cups so that each student has access to a stack.
3. Ask, "What measurements will you need to solve the problem?" (Students will need to measure the height of the entire first cup (h) and the part of the second cup that extends above the rim of the first cup when the cups are stacked (r).)
4. Ask students to write an equation for the height of the stack of two cups they have in front of them. Answer: $h + r$.
5. Now ask students, "What is the height of a stack of three cups?" Answer: $h + r + r$ or $h + 2r$.
6. Now ask students, "What is the height of a stack of ten cups?" Answer: $h + r + r + r + r + r + r + r + r + r$ or $h + 9r$.
7. Ask students to think about the pattern. Now ask, "What is the height of a stack of n cups?" Answer: $h + (n - 1)r$.
8. Ask students to write an equation for a stack of cups that would fit in a cabinet of the height you gave earlier. Answer: $h + (n - 1)r < C$.

Step	Complete the sentence
1. What is the problem asking?	The problem asks . . .
2. What facts are given in the problem. • What's the important information? • What's the unimportant information? • Why?	The important information is . . . The unimportant information is . . . because . . .
3. What do I need to do to solve the problem? • What operations will I need? • Write an equation.	To solve this problem, I will . . .
4. Solve the equation.	First I . . . Next I . . . Finally I . . .
5. Check the answer. • Is my answer correct?	When I substitute my answer for the variable . . .
6. Does the answer make sense? • Did I answer the question?	When I reread the problem, my answer . . .

9. Ask students to measure the dimensions of the cups and solve the equation for n.

10. Ask students, "Can n be a decimal? Why or why not? What must be done to solve the problem?" Answer: n must be a whole number of cups. The answer must be rounded down to a whole number.

11. Ask students why they must round down instead of up. Ask students what would happen if they rounded up.

12. Now ask students to check their answers by stacking the appropriate number of cups and seeing if they fit into the cabinet.

NOTE: If you don't have a file cabinet, you can ask students to write an equation for how old they are—in days and then in years.

Some Fun Math Activities The following activities may be helpful for kinesthetic learners.

Physical Number Line

Create a number line on the floor by moving desks back and using masking tape to "draw" a line. Mark the center "0" and then mark off positive and negative numbers on both sides of the center. Use the number line to work through the first few equations in Level 3 Book 1 dealing with signed numbers. Have students actually move along the number line to solve the problems.

Equations Flash Cards

Write the different parts of equations on flash cards and then have students physically move the cards around to show how to solve the equation. They can then create cards for intervening steps in the process.

Notes on Style

In this *Teacher's Guide,* the letter *T.* indicates Teacher. The letter *S.* indicates Student.

Some of the instructions for teaching the chapters are given in dialog form. Brackets [ ] are used to enclose the expected student response. Where necessary, parentheses ( ) are used to enclose what the teacher is expected to do but not to say in the dialog sections.

The pronouns *she* and *he* are used alternately between the skill books covered in this *Teacher's Guide*.

Level 3 Book 1: Signed Numbers

Introduction

If you have been referred to Book 1 by the diagnostic test, have S. take the pre-test to determine where to begin in this book. On the answer page to the pre-test is a list of the questions and their related chapter pages. When S. has missed a question, turn to the chapter to which the question refers and begin.

If S. is working through the complete series, there is no need to give the pre-test. The diagnostic test has already indicated a lack of skill, and it would be threatening and discouraging to have S. fail again at these questions. Simply begin with Chapter 1.

Chapter 1. Positive and Negative Numbers

Concepts:

- Understanding positive and negative numbers
- Recognizing and naming positive and negative numbers
- Knowing that zero is not positive or negative

Note: Some suggested dialogs between T. and S. are given. If S. doesn't need all or part of the suggested dialog, just remind her of relevant techniques.

Pages 7–9. Read the discussion slowly with S., stressing the bold lines. If either of you has difficulty seeing the bold lines, go over the lines with a marker or highlighter of another color. These lines represent points on the number line.

T: These are called *signed numbers* because they are written using positive or negative signs. When reading a signed number, we learn two things: how far from 0 we've gone on a number line, and the direction we've taken from 0. For example, –5 means 5 steps to the left of 0; 7 means 7 steps to the right of 0. Sometimes, these are called *directed numbers*. The formal mathematical name is *integers*.

Page 9. Read the directions to Exercise 1 with S. Remind S. to write the sign that identifies each number.

T: Let's look at these together. (Point to 0.) Let's start here. (Point to the –1 line.) What do we call this line or point? [S: –1] Yes.

Note: S. may call the numbers *negative* or *minus*.

T: (Point to the bold line.) What do we call this line or point? [S: –2] Good.

Have S. do problem 2 out loud, counting from 0. If S. succeeds, have her do the other problems, answering and checking each one against the answer key before going on.

Note: The worksheets from *Workbook for Level 3* provide extra practice. Assign them whenever you feel that S. needs this practice. Notes on using the worksheets are included under the title *Activity* throughout this *Teacher's Guide*.

Activity

Materials: Worksheet 1 and pencil

Procedure:

T: The first ten problems on this page are for practicing writing the numbers on a number line. The number lines go up and down in problems 1–8 and left and right in problems 9–15. A thermometer is an example of an up-and-down number line. On problems 1–3 and 9–15 we're going to write the number that represents each line or point. On 4–8 we'll write the number that represents the dark lines. Let's do problem 1 together. Remember, always start at 0. Let's count and write up from 0. What's the first line or point above 0? [S: +1] Yes.

Write:

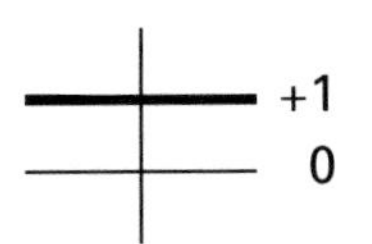

T: What's the line or point above +1? [S: +2] Yes. Write:

+2
+1
0

T: What's the line or point above +2? [S: +3] Yes. Write:

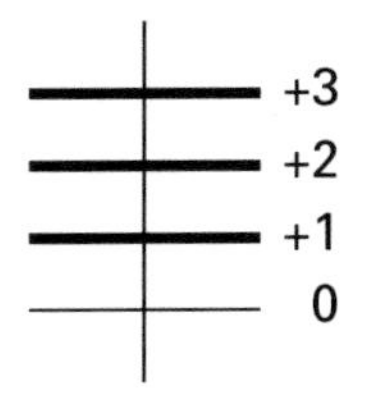

T: Now let's work below the 0. What's the first line or point below 0? [S: –1] Write:

+3
+2
+1
0
–1

T: What's the first line or point below –1? [S: –2] Write:

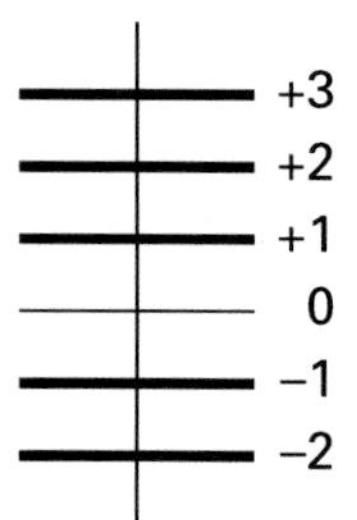

T: What's the first line or point below –2? [S: –3] Write:

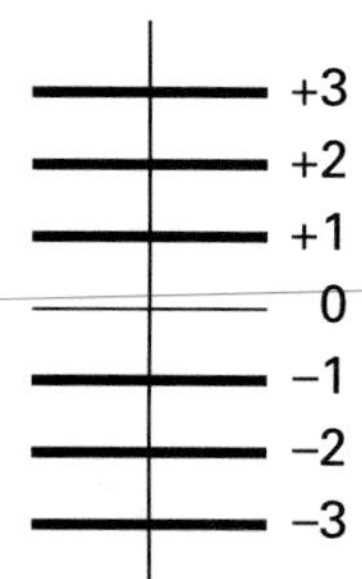

Remind S. to always start at 0 and work away from it one line or point at a time. Have S. do problems 2–3. Check each one as she does it. If she has any difficulty, help her.

T: Now let's do problem 4 together. On problems 4–8 we will only write the number that represents the dark lines. Remember, we start at 0. Let's count toward the dark line as I point. (Point to each line and have S. name the line. Write –3 when you get there.) [S: –1, –2, –3] Very good.

Have S. do problems 5–15. Check each one as she does it. If she has any difficulty, help her.

Chapter 2. Word Problems With Positive and Negative Numbers

Concepts:

- Understanding how to translate words into math
- Recognizing math in everyday life

Page 10. Go over some everyday examples of positive and negative numbers, for example, money, temperature, distance, etc. Read the three question prompts that are given to aid with problem solving, and ask questions to be sure S. understands them. Do the example, and answer the three problem-solving questions with the group.

Draw a thermometer to help in understanding the example problem. Have S. draw one at the same time. Explain that the drawings do not have to look the same; they are intended to help the S. translate the question from words to math. Point out that drawing a diagram is a good strategy for helping to figure out what a math problem is asking, especially for distance, height, volume, and quantity.

Page 11. Have S. do Exercise 2 using the same problem-solving process as in the example.

Chapter 3. Adding When the Signs Are the Same

Concepts:

- Knowing how to add positive numbers
- Knowing how to add negative numbers
- Knowing how to add positive and negative numbers on the number line

Pages 12–14. Read the discussion slowly with S., stressing the signs and the process. If S. has difficulty, try this additional explanation.

T: When adding numbers with like signs, we add the absolute values of the signed numbers and keep the same signs. When adding positives, we don't get over to the negative side and vice versa. This is because when adding, we follow the direction of the numbers. For example, to add –3 and –1:

1. Start at 0.
2. Move left 3 (–3).
3. Move left 1 (–1).
4. You're at –4.

Here is an example of adding negative sign numbers. (Read slowly to S.)

If the temperature is –5° and it drops a further 7°, the temperature will be –12°. We use a negative sign (–) to show a loss.

$$\begin{array}{r} -5^\circ \\ + \quad -7^\circ \\ \hline -12^\circ \end{array}$$

Page 15. Remind S. that each problem in Exercise 3 involves numbers with the same sign, so the answer will also have that sign, and that when there is no printed sign, the sign is positive. It is correct either to write or not to write the positive sign. Have S. do problem 2 and check it. If it is correct, have her do the other problems, answering and checking each against the answer key before going on. If addition is a problem, return to Level 1 Book 2.

Activity

Materials: Worksheet 2 and pencil

Procedure:

Note that the word *combine* is used to indicate addition or subtraction.

T: Let's do the first two together. (Point to problem 1.) How many steps did we take? [S: 3] Yes. What direction did we move both times? [S: Minus; or Negative.] Yes.

Write:
$$\begin{array}{r} -2 \\ + \ -1 \\ \hline -3 \end{array}$$

T: (Point to problem 2.) What is 3 and 2? [S: 5] Yes. What is the sign of the 5? [S: Plus; or None, because it's positive; or Positive.] Yes. Good.

Write:
$$\begin{array}{r} 3 \\ + \ 2 \\ \hline 5 \end{array}$$

Have S. do problem 3 and check it. If it is correct, have her do the other problems, answering and checking each against the answer key before going on. If addition is a problem, return to Level 1 Book 2.

Chapter 4. Multiple-Choice Questions

Concepts:

- Knowing how to answer multiple-choice questions
- Distinguishing the correct answer from wrong answer choices.

Page 16. Read the discussion. Ask for any questions or concerns that S. may have. Read the example and write it on the board. Do the math and then discuss the answer choices:

- 26 is the right number, but the wrong sign.
- –26 is the right number and the right sign.

Explain that after finding the right choice, there is no need to look at the other answer choices.

Explain that the wrong answers are often developed by following the correct answer steps but applying the common errors that people often commit. So, S. needs to be careful in doing computations.

Page 17. Read the instructions to Exercise 4. Have S. answer all questions by filling in the bubble next to the correct answer.

Chapter 5. Adding When the Signs Are Different

Concepts:

- Knowing to go to the right for positive numbers on a number line
- Knowing to go to the left for negative numbers on a number line
- Knowing, if the signs are different, to subtract the number closer to zero from the number farther from zero and to use the sign of the number farther from zero for the answer

Pages 18–20. Read the discussion with S., stressing counting right for positive and left for negative.

T: Now, to add unlike signs, first find the difference between the absolute values and then use the sign of the larger absolute value. Here is an example: –8 + 2. The difference between the absolute values is 6. Since 8 is the larger absolute value (even though –8 is the smaller number), use a minus sign in the answer: –6.

Show S. how it looks on a number line if necessary. The following example can clarify why –8 is not a larger number than +2: It's always better to have $2 (+2) in your pocket than to owe $8 (–8).

As you complete each example, have S. try working it on a separate piece of paper to be sure the skill has been learned before you continue. If it has not been learned, discuss and have S. try it again.

Page 21. Go over Exercise 5 as follows.

T: Let's do the first one together. Which number is farther from zero, 7 or 2? [S: 7]. What's the sign of the number farther from zero? [S: It's negative; or Minus.] Yes. So we write that sign for the answer.

Write:
$$\begin{array}{r} -7 \\ + \ 2 \\ \hline - \end{array}$$

T: Now we find the difference because the signs are different. What's 2 from 7? [S: 5] Yes.

Write:
$$\begin{array}{r} -7 \\ + \ 2 \\ \hline -5 \end{array}$$

T: If it's easier for you to find the difference if you rewrite the numbers so the number farther from zero is on top, then rewrite them.

Have S. do problem 2 and check it. If it is correct, have her do problems 3 to 6, answering and checking each against the answer key before going on. If addition is a problem, return to Level 1 Book 2.

Before S. does problems 7 and 8, explain that from now on, exercises will include word problems and multiple-choice questions. Review the process for these problems if necessary.

Activity

Materials: Worksheet 3 and pencil

Procedure:

T: Let's look at the first one together. Which number is farther from zero, 3 or 6? [S: 6] What's the sign of that number? [S: Positive; or Plus.] Yes. So we write that sign for the answer.

Write: $$\begin{array}{r} -3 \\ +\quad 6 \\ \hline + \end{array}$$

T: Now we find the difference. What's 3 from 6? [S: 3] Yes.

Write: $$\begin{array}{r} -3 \\ +\quad 6 \\ \hline 3 \end{array}$$

T: If it's easier for you to find the difference if you rewrite the numbers so the number farther from zero is on top, then rewrite them.

Write: $$\begin{array}{r} 6 \\ +\quad -3 \\ \hline 3 \end{array}$$

Point out that addition of opposite signed numbers always results in 0. For example, $-5 + 5 = 0$; $9 + (-9) = 0$.

Have S. do problem 2 and check it. If it is correct, have her do the other problems, answering and checking each against the answer key before going on. If subtraction is a problem, return to Level 1 Book 3.

Chapter 6. Adding More Than Two Signed Numbers

Concepts:

- Knowing to add all the positive numbers together
- Knowing to add all the negative numbers together
- Knowing how to add the positive and negative sums together
- Knowing that, when there are parentheses, you must combine what's inside first

Page 22. Read the discussion slowly with S.

T: Let's do the problem again together, slowly. First, let's list the positive numbers. What are they? [S: 5 and 6.] Yes. What is +5 and +6? [S: +11] Yes.

Write: $$\begin{array}{r} 5 \\ +\quad 6 \\ \hline 11 \end{array}$$

T: Now let's list the negative numbers. What are they? [S: –8, –9, and –1.] Yes.

Write: $$\begin{array}{r} -8 \\ -9 \\ +\quad -1 \\ \hline \end{array}$$

T: What sign will the answer to this be? [S: Negative; or Minus.] Yes. What is 8 and 9? [S: 17] What is 17 and 1? [S: 18] Good.

Write: $$\begin{array}{r} -8 \\ -9 \\ +\quad -1 \\ \hline -18 \end{array}$$

T: Now we take the sum of the positive numbers and add it to the sum of the negative numbers. So, what is +11 and –18? [S: –7] Very good. It's not so hard if you do it a little at a time like this.

Pages 23 & 24. Read up to Step 1 slowly. Work through Step 1 together. Point to the numbers printed in bold as you say the sum. For example:

T: 5 – 1 is 4. (Point to the bold 4.)

Read the rest of pages 23 and 24, slowly. As you complete each example, have S. try working it out on a separate piece of paper to be sure the skill has been learned before you continue. If it has not been learned, discuss and have S. try it again.

Page 25. Remind S. to be very careful in Exercise 6 to do what is inside parentheses first and to go slowly, working on paper when necessary, rather than in her head. (Whenever practical, be sure to encourage mental math, however.)

T: Let's do the first problem together. What do we do first? [S: Add all the positive numbers.] Yes. What are they? [S: 10 is the only one.] Good. And what do we do next? [S: Add all the negative numbers.] Yes. What are they? [S: –5, –5, and 10.] Good.

Write: $$\begin{array}{r} -5 \\ -5 \\ +\quad -10 \\ \hline \end{array}$$

T: What is the sign of this answer? [S: Negative.] Yes. What is 5 and 5 and 10? [S: 20] Good.

Write: $$\begin{array}{r} -5 \\ -5 \\ +\quad -10 \\ \hline -\,20 \end{array}$$

T: Remember the positive 10 we had? We add that to –20. What do we get? [S: –10] Very good.

Then have S. do problem 2 and check it. If it is correct, have her do the other problems, answering and checking each against the answer key before going on. If addition is a problem, return to Level 1 Book 2. If subtraction is a problem, return to Level 1 Book 3.

Activity

Materials: Worksheet 4 and pencil

Procedure:

T: Let's do the last problem together this time. What is 4 – 5? [S: –1] Yes.

Write: (4 – 5)
(–1)

T: What is 5 – 9? [S: –4] Yes.

Write: (4 – 5) + (5 – 9)
(–1) (–4)

T: What is 6 – 5? [S: 1] Yes.

Write: (4 – 5) + (5 – 9) + (6 – 5)
(–1) (–4) (1)

T: What is 3 + 1? [S: 4] Yes.

Write: (4 – 5) + (5 – 9) + (6 – 5) + (3 + 1) =
(–1) (–4) (1) (4)

T: Now we'll add the positive numbers. What is 1 + 4? [S: 5] Yes. And now we'll add the negative numbers. What is –1 and –4? [S: –5]. Yes. So what is +5 and –5? [S: 0] Great!

Write: (4 – 5) + (5 – 9) + (6 – 5) + (3 + 1) = 0
(–1) (–4) (1) (4)

T: Now the others will seem easy.

Have S. do the first problem and check it. If it is correct, have her do the other problems, answering and checking each against the answer key before going on.

Chapter 7. Subtracting Signed Numbers

Concepts:

- Knowing that we subtract to find differences
- Knowing that we subtract the starting number from the ending number
- Knowing that we add the opposite of the second number

Pages 26–29. Read the discussion slowly with S., pointing to and counting with the bold indicators and stressing the chapter concepts. You can relate the process to other examples of change, such as finding the mileage of a trip or how many dollars were deposited in an account:

end odometer reading	end balance
– start odometer reading	– start balance
number of miles traveled	$ deposited

Stress the importance of order in subtraction and division problems. Remind S. that $\frac{4}{8}$ does not equal $\frac{8}{4}$ and that 7 – 2 does not equal 2 – 7.

Complete Examples 1–6, and then go over the chapter concepts again. If S. has difficulty, try this additional explanation.

T: When adding a number using a number line, we move in the direction of the signed number. Subtraction, which is the inverse, or opposite, of addition, tells us to move in the opposite direction of the signed number. For example, –3 – (2):

1. Move left 3 (–3).
2. Move 2 steps in the opposite direction of + (–2).
3. You're at –5.

The rule to use when subtracting without using a number line is: To subtract a number, add its opposite. (For example, –5 and 5 are opposites since they are the same number of steps from 0 in opposite directions.) Remember that only the number being taken away is replaced by its opposite.

Page 29. Read the directions to Exercise 7 with S.

T: Let's work on the first two together. First, we'll just make the sign changes.

Write: (8) – (4) =
(8) + (–4) =

T: (Point to the + between the two sets of parentheses.) This + tells you to add. (Point to the – in front of the 4.) We changed the + to a –. What is 8 minus 4? [S: 4] Very good.

Write: (–4) – (9) =

T: (Point to the – between (–4) and (9).) What do I do with this –? [S: Change it to +.] Yes. What do I do with the (9)? [S: Change it to (–9).] Yes.

Write: (–4) + (–9) =

T: What is –4 and –9? [S: –13] Very good.

Have S. do problem 3 and check it. If it is correct, have her do the other problems, answering and checking each against the answer key before going on. If signs are a problem, return to Level 3 Book 1, Table of Contents to locate the type of sign-combining that is a problem. Review the necessary skills with S. before going on.

Activity

Materials: Worksheet 5 and pencil

Procedure:

T: Remember that you add the opposite of the second number. What will we rewrite for this first problem? [S: 4 + (–5)] Yes. Very good.

Write: (4) + (–5) =

T: What is 4 and –5? [S: –1] Very good.

Write: (4) + (–5) = –1

Have S. do problem 2 and check it. If it is correct, have her do the other problems, answering and checking each against the answer key before going on. If signs are a problem, return to Level 3 Book 1, Table of Contents to locate the type of sign-combining that is a problem. Review the necessary skills with S. before going on.

Chapter 8. Subtracting More Than Two Signed Numbers

Concepts:

- Knowing that when we see a subtraction sign between parentheses, we change the sign of the number and add
- Knowing how to add signed numbers
- Knowing how to subtract signed numbers

Pages 30 & 31. Read the discussion slowly with S., stressing the sign changes and the rule that tells when and what to do. As you complete each example, have S. try working it on a separate piece of paper to be sure the skill has been learned before you continue. If it has not been learned, discuss and have S. try it again.

Page 32. Look at Exercise 8 with S. Point out that parentheses are either used to separate numbers, as in problems 1–4, or to point out work that must be done first, as in problems 5–7. Remind S. to check parentheses for priority work. Simplification within parentheses must be done before any sign-changing.

Now work the problems in Exercise 8.

T: Let's do the first problem together. Do you see the three sets of parentheses? [S: Yes.] Between them are subtraction signs. What do we do with these subtraction signs? [S: Change them to addition signs.] Yes. What do we do with the numbers in parentheses after the subtraction signs? [S: Change them to the opposite signs.] Yes.

Write: (–7) + (–4) + (+2) =

T: Now what do we do? [S: Add the positive numbers.] Yes. What are they? [S: Only 2.] Yes. What do we do next? [S: Add the negative numbers.] Yes. What are they? [S: –7 and –4] What is –7 and –4? [S: –11] Yes. What is 2 and –11? [S: –9] Yes. Very good.

Have S. do problem 2 and check it. If it is correct, have her do problems 3 and 4, answering and checking each against the answer key. Do problem 5 together to be sure S. does priority work in parentheses first. If signs are a problem, return to Level 3 Book 1, Table of Contents to locate the type of sign-combining that is a problem. Review the necessary skills with S. before going on.

Activity

Material: Worksheet 6 and pencil

Procedure:

T: Let's do the last one together this time. There are three sets of parentheses. In each set of parentheses are two numbers to add or subtract. Let's do what is inside each set first. (Point to the first set of parentheses.) What is –14 + 10? [S: –4] Yes.

Write: (–14 + 10) – (100 – 120) – (–35 + 5) = (–4)

T: What is 100 – 120? [S: –20] Yes.

Write: (–14 + 10) – (100 – 120) – (–35 + 5) =
(–4) – (–20)

T: What is –35 + 5? [S: –30] Yes.

Write: (–14 + 10) – (100 – 120) – (–35 + 5) =
(–4) – (–20) – (– 30) =

T: Now we add the opposite of the second and third numbers.

Write: (–4) + (+20) + (+30) =

(Check that S. understands these changes.)

T: Now we add all the positive numbers. What is 20 + 30? [S: 50] Yes. Then we add the negative numbers. What is the negative number? [S: –4] Yes. Since that's the only negative, we add the 50 and the –4.

Write: (50) + (–4) =

T: What is 50 and –4? [S: 46] Very good.

Write: (50) + (–4) = 46

Have S. do problems 1–9, answering and checking each against the answer key. If they are correct, have her do the other problems, answering and checking each before going on. If signs are a problem, return to Level 3 Book 1, Table of Contents to locate the type of sign-combining that is a problem. Review the necessary skills with S. before going on.

Chapter 9. Multiplying Signed Numbers

Concepts:

- Knowing that in multiplying signed numbers the answer is positive when the signs are the same

- Knowing that in multiplying signed numbers the answer is negative when the signs are different
- Knowing to multiply when a signed number, variable, or another set of parentheses is right next to parentheses

Pages 33 & 34. Some weight-change examples may help explain the rules before you read the discussion.

1. Sam started a bodybuilding program. He plans to gain 3 (+3) pounds a week for 4 (+4) weeks. How much weight will he gain?

   Answer: $(+3)(+4) = +12$ pounds

2. Sam started a diet 2 weeks ago (–2). He lost 3 pounds (–3) each week. How much has he lost?

   Answer: $(-2)(-3) = +6$ pounds

3. Sam wants to diet for 3 weeks (+3), losing 1 pound (–1) each week. What is his weight change in 3 weeks?

   Answer: $(+3)(-1) = -3$ pounds

Now read the discussion slowly with S., stressing the chapter concepts above before doing each example. As you complete an example, have S. try working it on a separate piece of paper to be sure the skill has been learned before you continue. If it has not been learned, discuss and have S. try it again.

Page 34. Read the directions to Exercise 9 with S. Remind S. that parentheses tell her to multiply. Ask S. to go through problems 1–8 with you, just telling you if the signs are the same or different, and if the answer will be positive or negative. Remind S., if necessary, that if the signs are the same, the answer is positive, and if the signs are different, the answer is negative.

1. different: negative
2. same: positive
3. different: negative
4. same: positive
5. different: negative
6. different: negative
7. same: positive
8. different: negative

If S. has any difficulty, repeat this approach several times. Now have S. complete the problems, answering and checking each against the answer key before going on. If multiplying is a problem, return to Level 1 Book 4.

Activity

Materials: Worksheet 7 and pencil

Procedure:

Remind S. that parentheses tell her to multiply. Ask S. to go through problems 1–10 with you, just telling you if the signs are the same or different, and if the answer will be positive or negative.

1. same: positive
2. same: positive
3. same: positive
4. same: positive
5. same: positive
6. different: negative
7. different: negative
8. different: negative
9. different: negative
10. different: negative

Have S. do problem 1 completely and check it. If it is correct, have her do the other problems, answering and checking each against the answer key before going on. If multiplying is a problem, return to Level 1 Book 4.

Chapter 10. Multiplying More Than Two Signed Numbers

Concepts:

- Knowing that in multiplying two signed numbers the answer is positive when the signs are the same
- Knowing that in multiplying two signed numbers the answer is negative when the signs are different
- Knowing to multiply from left to right, starting with two numbers, then multiplying those numbers and the next number, then that result and the next number, etc.

Pages 35 & 36. Read the discussion slowly with S., stressing the order indicated in bold. As you complete each example, have S. try working it on a separate piece of paper to be sure the skill has been learned before you continue. If it has not been learned, discuss and have S. try it again.

Page 37. Read the directions to Exercise 10 with S.

Write:

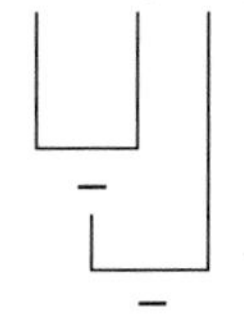

Write: $(-6)(4)(1) = -24$

Do problem 5 together. Remind S. to do what is inside the parentheses first.

Write: $(5 - 2)(-3)(3 + 1) =$

$(3) \quad (-3) \quad (4) \quad =$

Write: (+) (–) (+)

–

–

So: $(3)(-3)(4) =$

$(-9)(4) =$

-36

Now have S. do problem 2 and check it. If it is correct, have her do the other problems, answering and checking each against the answer key before going on. If multiplying is a problem, return to Level 1 Book 4. If multiplying signs is a problem, return to Chapter 9 in the skill book and this chapter in this *Teacher's Guide* and review the skill with S. before going on.

Activity

Materials: Worksheet 8 and pencil

Procedure:

Remind S. to simplify what is inside the parentheses first, and then do the multiplying. As S. multiplies, be sure she is keeping track of the signs. If necessary, have her do the signs before multiplying. Remind S. to write something down when necessary in order to figure it out. Have S. do problem 1 and check it. If it is correct, have her do problems 2–19, answering and checking each against the answer key before going on. Work problem 20 together. If you need to illustrate the rule that 0 times any number is 0, use the following example.

T: If you have no money ($0) and I offer to give you 5 times what you have, how much will I give you? [S: Nothing; or $0.] T: So 5 times nothing (0) equals nothing (0).

If multiplying is a problem in completing Worksheet 8, return to Level 1 Book 4. If multiplying signed numbers is a problem, return to Chapters 9 and 10, Level 3 Book 1.

Chapter 11. Dividing Signed Numbers

Concepts:

- Knowing that in dividing signed numbers the answer is positive if the signs are the same
- Knowing that in dividing signed numbers the answer is negative if the signs are different

Note: The above concepts show that, since division and multiplication are inverse operations, we can use the same sign rule for both. If S. has difficulty with this idea, use the following examples to illustrate.

1. Rosa owes $8 (–8) for her dinner. She borrows from 4 friends (4). How much does she owe to each friend?

   Answer: $\frac{-8}{4} = -2$ $2 owed to each friend

2. Rosa paid $20 (20) for pizza. She split the cost with 3 friends (4). How much did each friend pay?

   Answer: $\frac{20}{4} = 5$ $5 paid by each person

Pages 38 & 39. Read the discussion only with S., stressing the chapter concepts above before doing each example. Remind S. that a horizontal bar separating numbers means several things. It can mean a fraction (how many parts out of a total number of parts); it can mean division (the top number divided by the bottom number); or it can mean ratio (a comparison of two numbers).

Point out that zero can be divided by a non-zero number, but the result will always be zero. Mathematically, it is not possible to divide a non-zero number by zero. For example, $\frac{0}{7} = 0$; $\frac{7}{0}$ is not allowed.

Page 40. Read the directions to Exercise 11 with S. Ask S. just to tell you if the signs are the same or different for problems 1–4. Remind S. that if the signs are the same, the answer is positive and if the signs are different, the answer is negative (just like multiplying signed numbers).

1. different: negative
2. different: negative
3. same: positive
4. same: positive

Have S. do the first four problems and check them. If they are correct, have her do the other problems, answering and checking each against the answer key before going on. If combining the numbers (such as 5 – 10 in problem 5) presents difficulty, return to Chapters 3, 5, and/or 7 before going on. If dividing is a problem, return to Level 1 Book 5.

Activity

Material: Worksheet 9 and pencil

Procedure:

Have S. do problem 6 and check it. If it is correct, have her do the other problems, answering and checking each against the answer key before going on. If combining the numbers is a problem, return to Chapters 3, 5, and/or 7 before going on. If dividing is a problem, return to Level 1 Book 5.

Chapter 12. Using More Than One Operation to Find an Answer

Concepts:

- Knowing that the four operations are adding, subtracting, multiplying, and dividing
- Knowing the order of operations:
  1. Combine inside parentheses first
  2. multiply and/or divide next
  3. add and/or subtract last
  4. divide again, if necessary

Page 41. Read the discussion slowly with S. up to Step 1.

T: First we work inside the parentheses. Then we multiply and divide. And, last, we add and subtract.

Reassure S. that with practice this sequence will come naturally. Remind S. that the order of operations just makes sure that everyone performs mathematical steps in the same order so that everyone can end up with the same answer. We read math from left to right, just as we read words. At the beginning of each new operation, go to the beginning of the new sentence.

Read and work each step of the problem slowly. In Step 3, be sure that S. simplifies the dividend and divisor before attempting to divide. Have S. try it on a piece of paper before going on. Tell S. what procedure to use for each step as written on the page. Write the problem again for S. and let her try it alone. If she is successful, go on to page 42. If not, try writing the work for her like this, one line at a time.

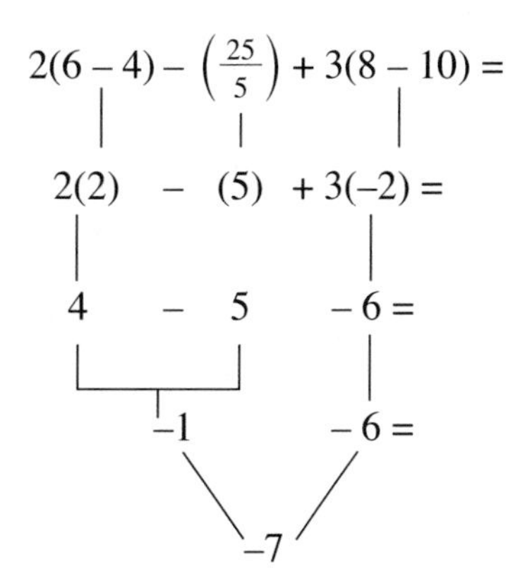

Write this problem again and have S. try it again. Stay with this until S.'s answer is correct and she is confident.

Pages 42 & 43. Read and work the examples slowly with S. As you complete each example, have S. try working it on a separate piece of paper to be sure the skill has been learned before you continue. If it has not been learned, discuss and have S. try it again.

Page 44. Have S. do problem 1 of Exercise 12 and check it. If it is correct, have her do the problems, answering and checking each against the answer key before going on. If addition is a problem, return to Level 1 Book 2. If subtraction is a problem, return to Level 1 Book 3. If multiplication is a problem, return to Level 1 Book 4. If division is a problem, return to Level 1 Book 5. If signs are a problem, return to Level 3 Book 1, Table of Contents to locate the type of sign-combining that is a problem. Review the necessary skills with S. before going on.

Activity

Materials: Worksheet 10 and pencil

Procedure:

Use problem 15 as an example. Write the work like this, one line at a time.

$$2\left(\frac{14}{7}\right) + 7\left(\frac{24}{-6}\right) - 3(-5 + 3) =$$

$$2(2) + 7(-4) - 3(-2) =$$

$$4 - 28 + 6 =$$

$$-24 \quad +6$$

$$-18$$

Then, have S. try it. If she is successful and confident, have her do problem 1 and check it. If it is correct, have her do the other problems, answering and checking each against the answer key before going on. If addition is a problem, return to Level 1 Book 2. If subtraction is a problem, return to Level 1 Book 3. If multiplication is a problem, return to Level 1 Book 4. If division is a problem, return to Level 1 Book 5. If signs are a problem, return to Level 3 Book 1, Table of Contents to locate the type of sign-combining that is a problem. Review the necessary skills with S. before going on.

Post-Test. After you have completed this chapter, administer the post-test. Review the appropriate pages for any incorrect answers. Then retest. When S. has successfully completed the work, go on to the next book.

Level 3 Book 2: Solving Equations

Introduction

If you have been referred to this book by the diagnostic test, have S. take the pre-test to determine where to begin in this book. On the answer page to the pre-test is a list of the questions and their related chapter pages. When S. has missed a question, turn to the chapter referred to and begin.

If S. is working through the complete series, there is no need to give the pre-test. The diagnostic test has already indicated a lack of skill, and it would be threatening and discouraging to have S. fail again at these questions. Simply begin with Chapter 1.

Chapter 1. Variables

Concepts:

- Understanding the use of a variable (letter such as x) to represent a specific, but unknown, number
- Knowing the following symbols for operations:

  Addition: +

  Subtraction: –

  Multiplication: (3)(4)
  $3(x)$
  $3x$

  Division: $\frac{2}{3}$
- Knowing the following clue words for operations:

  Addition: *more than, the sum of*

  Subtraction: *decreased by, less than, the difference between*

  Multiplication: *times*

  Division: *divided by*

Pages 7 & 8. Tell S. that you will read this section together and that he shouldn't worry about memorizing the large amount of information here, since it will be presented a little at a time. Read the discussion slowly with S., stressing the chapter concepts. Reassure S., if he feels overwhelmed, that you will work slowly.

Point out that in algebra the symbol × is not used to indicate multiplication. x is only for use as a variable. Parentheses are used where necessary. Point out also that a "naked" variable is a variable with a coefficient of positive one: $x = 1x$.

Page 8. Read Example 1, *Expression in words.*

T: The clue words here are *more than. More than* tells us to add. Usually we write the letter first, and then add the number. $5 + x$ is not wrong, but $x + 5$ is the way we usually write it.

(Read Example 2, *Expression in words.*) The clue words here are *less than. Less than* tells us to subtract. Remember that the order of the terms is important in subtraction. We must write "5 less than x" as "$x - 5$," not "$5 - x$." Otherwise, it doesn't say "5 less than x," but "x less than 5."

Let's think about this with numbers. If we were saying "5 less than 10," we'd write: (Write: $10 - 5$). So, 5 less than x is: (Write: $x - 5$).

Read the other examples slowly with S. Point to the algebraic expressions as you read. Then, return to Example 1 and cover the algebraic expression. Read the words for the example and have S. try to write the algebraic expression. Do this with all 12 examples, uncovering the answer and comparing S.'s written work. If S. has difficulty writing, help him. Reverse the procedure, covering the expression in words and having S. say the words. If any are wrong, have S. copy the problems and answers, or copy them for him, to do for practice at home.

Page 9. Read the directions to Exercise 1-A with S. Read the *Expression in words* slowly with S. Have S. copy the problems and answers to do for practice at home.

If S. has difficulty writing, have him say the expressions in words for Exercise 1-B and write for him. Have S. copy the problems and answers to do for practice at home.

Activity

Materials: Worksheet 11 and pencil

Procedure:

Read the *Expression in words* with S. and have him try to find the algebraic expression for problems 1–9. If he succeeds, have him do the other problems, answering and checking each against the answer key. Have S. copy the problems and answers to do for practice at home.

Chapter 2. Word Problems With Constants and Variables

Concepts:

- Understanding how to translate word problems into math equations
- Recognizing math in everyday life

Page 10. Go over some everyday examples of constant and variable numbers. Constants are numbers that never change, for example, a difference in age, the number of chapters in a book, anything that is written as a number in the question. Variables are values that change depending on what you are being asked to find, for example, a sale price, someone's age, the number of miles a car has been driven. Read the three problem-solving questions and ask about meaning and understanding. Do the example and answer the three problem-solving questions with the group.

Page 11. Have S. use the three steps to do Exercise 2.

Chapter 3. Evaluating Algebraic Expressions

Concepts:

- Knowing what the operation symbols tell you to do
- Knowing how to substitute numbers given into the algebraic expression
- Knowing how to evaluate the algebraic expression using the values given for the letters

Pages 12 & 13. Read the discussion slowly with S., stressing the replacement of the variables (letters) with the given values (numbers), pointing to the letters and numbers as you read. Tell S. this process is important for using formulas and checking solutions of equations. It is important, when first replacing a variable with its given value, to put parentheses around the number until you decide what operations to use. For example, in finding $2x$ if $x = -3$, start by saying to yourself "$2x$ means 2 times –3." This leads you correctly to $2(-3) = -6$ rather than incorrectly to $2 - 3 = -1$.

Page 13. Read the directions for Exercise 3 with S. Do problem 1 together.

T: $\frac{r}{t}$ means r divided by t. When we substitute values for r and t, we get 3 divided by 9. Writing this algebraically looks like a fraction $\frac{3}{9}$. We can reduce the quotient (result of division) just as with fractions, by dividing common factors into both terms.

Write: $\frac{r}{t} = \frac{3}{9} = \frac{1}{3}$

Have S. do problem 2 and check it. If it is correct, have him do the other problems, answering and checking each against the answer key before going on. Point out that from now on, exercises will include some problems expressed in words, such as problem 10. If fractions are a problem, return to Level 2 Book 1 and use the pre-test to locate the chapters to teach before going on in Book 2.

Activity

Materials: Worksheet 12 and pencil

Procedure:

For problem 1, write: $st = 2(5) = 10$

Problem 20:

Write: $\frac{z}{x} = \frac{5}{3}$

T: We always want to write out answers in simplest terms. It is fine, however, to leave an improper fraction in algebra, e.g., $\frac{5}{3}$ rather than $1\frac{2}{3}$. Improper fractions are easier to check in equation problems.

Have S. do problem 2 and check it. If it is correct, have him do the other problems, answering and checking each against the answer key before going on. If fractions are a problem, return to Level 2 Book 1 and use the pre-test to locate the chapters to teach before going on in Book 2.

Chapter 4. Using More Than One Operation to Evaluate Algebraic Expressions

Concepts:

- Knowing the order of operations
- Knowing how to evaluate algebraic expressions

Pages 14 & 15. Review with S. the meaning of the phrase *order of operations*. Then read the discussion slowly with S., stressing combining numbers within parentheses and the other operations.

Page 15. Read the directions to Exercise 4 with S. Have S. do problem 1 and check it. If it is correct, have him do the other problems, answering and checking each against the answer key before going on. If necessary, do the last problem together. Simplify above the division bar, below the division bar, and then divide.

Write: $\frac{3d+2d}{d} =$

$$\frac{3(-3)+2(-3)}{-3}$$

$$\frac{-9+(-6)}{-3}$$

$$\frac{-9-6}{-3}$$

Write: $\frac{-15}{-3} = 5$

Have S. copy the directions and problems (or copy for him if reading or writing is a problem) to practice at home.

Activity

Materials: Worksheet 13, list of operations from page 14 of Book 2, and pencil

Procedure:

Write: $r = 3$ and $s = 4$

$2(r + s) =$

$2(3 + 4) =$

$2(7) =$

14

Have S. do problem 2 and check it. If it is correct, have him do the other problems, answering and checking each against the answer key before going on. The answer to problem 20 is 1. Remind S. that a number can be divided by itself. The quotient is always 1 ($\frac{15}{15} = 1$).

If S. wants a copy of these to practice at home, encourage it. Be sure he also has a copy of the list of operations.

Chapter 5. Multiple-Choice Questions

Concepts:

- Understanding how to answer multiple-choice questions
- Distinguishing the correct answer from wrong answer choices

Pages 16 & 17. Read the discussion. Ask for any questions or concerns that S. may have. Read the example and write it on the board: $3r(t + s)$ when $r = 4$, $t = 6$, and $s = -2$

Do the math and discuss the answer choices:

- 48 is the right number and the right sign.
- –48 is the right number, but the wrong sign.

Explain that after finding the right choice, there is no need to look at any other answers.

To get answer d:

$3r(t + s)$

$3(4)(6 + (-2)) =$

$3(4)(-4) =$

$12(-4) = -48$

The mistake was made in adding 6 and –2 and getting –4, instead of 4. Explain that the wrong answers are often developed by following the correct answer steps but applying common errors that people often make. So, S. needs to be careful in doing computations.

Page 18. Read the instructions to Exercise 5. Have S. answer all questions by filling in the bubble for the letter that represents the correct answer.

Chapter 6. Solving Equations With One Inverse Operation

Concepts:

- Understanding the equality (balance) of equations
- Knowing that you must maintain the equality (balance) of an equation
- Knowing that to "get rid" of something you do the inverse (opposite) operation to both sides
- Knowing the inverse operations:

Operation	Inverse of that operation
+	–
–	+
×	÷
÷	×

Show S. the *Operation/Inverse of that operation* chart above.

T: Remember how you check subtraction by addition and check division by multiplication? That's because they are inverse operations: they "undo" each other.

Pages 19 & 20. Read the discussion slowly with S., stressing the chapter concepts above. If S. has difficulty, try adding the following examples and explanation.

Write: $x + 3 - 10$

T: We want to end up with $(x + 0)$ or (x). In $(x + 3)$ we are adding 3 to the x. We add –3 to both sides. We subtract from both sides to keep the balance.

Write:

$$\begin{array}{rcl} x + 3 &=& 10 \\ -3 && -3 \\ \hline x + 0 &=& 7 \\ x &=& 7 \end{array}$$

(S. can omit "+ 0" step if it is understood.)

Write:

$$\begin{array}{rcl} m + 5 &=& 20 \\ -5 && -5 \\ \hline m + 0 &=& 15 \\ m &=& 15 \end{array}$$

(S. can omit "+ 0" step if it is understood.)

T: To get m alone we add the opposite of +5, that is, –5, to both sides.

Write:

$$\begin{array}{rcl} s - 2 &=& 6 \\ +2 && +2 \\ \hline s + 0 &=& 8 \\ s &=& 8 \end{array}$$

(S. can omit "+ 0" step if it is understood.)

T: To get s alone we add the opposite of –2, that is, +2, to both sides.

In general, ask yourself, "What operations are happening to the variable?" Then use the inverse to undo that operation.

Pages 21–23. Review the *Operation/Inverse of that operation* chart before doing page 21. Read the discussion slowly with S. As you complete each example, have S. try working it on a separate piece of paper to be sure the skill has been learned before you continue. If it has not been learned, discuss and have S. try it again.

If S. still has difficulty, try this additional explanation, adding your own examples if necessary.

T: The goal of solving an equation is to find the value of a single positive variable. First look at the variable to see what operations are being applied to it. Then undo those operations.

Since adding a number and its opposite always equals zero, if we have a number added to a variable, we can add that number's opposite to get the variable itself. This process is called the *additive inverse*. In using it, we have to add the same number to each side of the equation in order to keep the balance.

Note: The additive inverse is the opposite of a signed number. When you add a signed number and its inverse, you always get 0. For example, 8 is the additive inverse of –8; 8 + (–8) = 0.

T: When a variable is being multiplied or divided by a number, we can use the *multiplicative inverse*. This process is based on the idea that multiplication and division are opposite operations and on the idea that any number divided by itself equals positive one. In solving an equation, remember to use the multiplicative inverse on each side of the equation.

Note: The multiplicative inverse is also called the reciprocal. When you multiply a number and its inverse, you always get 1. For example, $\frac{1}{5}$ is the multiplicative inverse of 5 $\left(\frac{5}{1}\right)$; $\left(\frac{1}{5}\right)\left(\frac{5}{1}\right) = 1$.

T: Both sides of your final equation should be exactly equal. If they're not, go back and check your equation work. The error may be of signs or simple miscalculation. It's nothing to get frustrated about. Sometimes, it's even a small mistake in the check itself.

Page 23. Read the directions to Exercise 6 with S. Write: $3x = -21$

T: The x is being multiplied by 3, so we will divide by 3 on both sides.

Write: $\frac{3x}{3} = \frac{-21}{3}$

$$\frac{{}^{1}\cancel{3}x}{{}_{1}\cancel{3}} = \frac{-\cancel{21}^{-7}}{\cancel{3}_{1}}$$

$$\frac{x}{1} = \frac{-7}{1}$$

$$x = -7$$

If division and equivalent numbers are a problem, return to Level 2 Book 1 to review the parts that are confusing. Have S. do problem 2 and check it. If it is correct, have him do the other problems, answering and checking each against the answer key before going on.

Activity

Materials: Worksheet 14 and pencil

Procedure:

Do the first problem together.

Write: $x + 3 = 8$

T: To get x alone, we add the opposite of +3, that is, –3, to both sides.

Write:
$$\begin{aligned} x + 3 &= 8 \\ -3 \;\; & -3 \quad \text{(additive inverse)} \\ \hline x + 0 &= 5 \\ x &= 5 \end{aligned}$$

Have S. do problem 2 and check it. If it is correct, have him do the other problems, answering and checking each against the answer key before going on.

Chapter 7. Solving Equations With Two Inverse Operations

Concepts:

- Knowing to add or subtract first
- Knowing to multiply or divide next
- Knowing how to check an answer

Pages 24 & 25. Read the discussion slowly with S., stressing the first two chapter concepts listed above.

As you complete each example, have S. try working it on a separate piece of paper to be sure the skill has been learned before you continue. If it has not been learned, discuss and have S. try it again.

Page 26. Read the directions to Exercise 7 with S.

$$\begin{aligned} 3x + 9 &= 16 \\ -9 \;\; & -9 \\ \hline 3x \quad &= 7 \end{aligned}$$

T: Use the inverse of addition (additive inverse) to add –9 to both sides.

$$\frac{3x}{3} = \frac{7}{3}$$

T: Use the inverse of multiplication (multiplicative inverse) to divide both sides by 3.

$$x = \frac{7}{3}$$

T: This is an acceptable, easily checked answer.

Check the answer.

Write: $3x + 9 = 16$

$$3\left(\frac{7}{3}\right) + 9 = 16$$

$${}^{1}\cancel{3}\left(\frac{7}{\cancel{3}_{1}}\right) + 9 = 16 \quad \text{(Divide out a common factor of 3.)}$$

$$7 + 9 = 16$$

$$16 = 16 \text{ check!}$$

Point out to S. that the answer is correct when the last line of the check is the same on both sides. Suggest that S. check each problem by substituting the solution for the variable in this way. Tell S. that from now on exercises will include multiple-choice problems. Remind S. to fill in the bubble for the correct answer.

Activity

Materials: Worksheet 15 and pencil

Procedure:

Write: $2x + 3 = 13$

$$\begin{aligned} 2x + 3 &= 13 \\ -3 \quad -3 & \qquad \text{(additive inverse)} \\ 2x + 0 &= 10 \\ \frac{2x}{2} &= \frac{10}{2} \\ x &= 5 \end{aligned}$$

Check the answer.

Write: $2x + 3 = 13$

$$\begin{aligned} 2(5) + 3 &= 13 \\ 10 + 3 &= 13 \\ 13 &= 13 \text{ check!} \end{aligned}$$

T: If it doesn't check, start over. The error may be of signs or simple miscalculation. It's nothing to get frustrated about. Sometimes, it's even a small mistake in the check itself.

Have S. do problem 2 and check it. If it is correct, have him do the other problems, answering and checking each against the answer key before going on.

Chapter 8. Combining Like Variables

Concepts:

- Recognizing *like variables*
- Knowing to combine (add or subtract) *like variables*
- Knowing how to combine (add or subtract) *like variables*
- Knowing that a "naked" variable is a variable with a coefficient of one, i.e., that x is really $1x$
- Knowing that parentheses tell you to multiply
- Knowing how to multiply numbers and variables

Pages 27 & 28. Read the discussion slowly with S., stressing the chapter concepts. Remind S. that *combining* means adding or subtracting. So, *combining like variables* means adding or subtracting the numbers of the letters that are alike. Variables aren't changed when they're combined, e.g., $3x + 5x = 8x$, not $8x^2$.

T: If we have 3 apples and 2 birds, can we combine them into a single item? [S. No.] Well, that's the way it is with variables. If we think of the 3 apples as $3a$ and the 2 birds as $2b$, we can't combine them into a single item. We could also have 4 apples and eat one apple. That would be a subtraction. We'd have 3 apples left. (Write: $4a - a = 3a$)

Note: The name for the number in front of the variable is *numerical coefficient*. It is a factor (multiplier) of the term, just as the variable is a factor. It tells how many of the variables there are: $5x$ means $x + x + x + x + x$. Factors can be multiplied in any order: $3(6) = 6(3)$.

Page 29. Read the directions to Exercises 8-A and 8-B with S. Discuss each line carefully. Have S. do problem 1 of Exercise 8-B and check it. If it is correct, have him do the other problems, answering and checking each against the answer key before going on. If signs are a problem, return to Level 3 Book 1, Table of Contents to locate the type of sign-combining that is a problem. Review the necessary skills with S. before going on. Have S. copy the problems and answers to do for practice at home.

Activity

Materials: Worksheets 16-A and 16-B and pencil

Procedure:

Have S. do problems 1–10 on Worksheet 16-A and check his work. If it is correct, have him do the other problems, answering and checking each against the answer key before going on. If S. has difficulty, review Chapter 8.

Now look at Worksheet 16-B.

T: If we wanted to simplify $2(3 + 4)$, we would combine 3 and 4, and then multiply the sum by 2, that is, $2(7) = 14$. This follows the order of operations. If you have a variable and a number inside the parentheses, multiply: $2(x + 4) = 2(x) + 2(4) = 2x + 8$.

It's important to remember in using this multiplication process to multiply each and every term inside the parentheses by the number outside the parentheses. For example:

1. $2(x + 4)$ does not equal $2x + 4$.
2. $-2(3y - 1) = (-2)(3y) + (-2)(-1)$
   $= -6y + 2$

Chapter 9. Combining Variables to Solve Equations

Concepts:

- Knowing to combine variables first
- Knowing to use inverse operations next
- Knowing how to check answers

- Knowing to multiply both terms in the parentheses by the term outside (with its sign), i.e., *distributing*.

Note: *Combining like terms* is applied only to terms on the same side of the equals sign. To shift terms from one side to the other, you have to use the additive inverse. Once the like terms are combined (on one or both sides of the equation), the equation can be solved using the additive inverse and multiplicative inverse as in preceding chapters.

Note: *Distributing* means using the *distributive property*. This means that the product of a number and the sum of two or more other numbers is equal to the sum of the individual products. For example, $2(x + y) = 2x + 2y$ and $2(a + b - c) = 2a + 2b - 2c$.

Page 30. Read the discussion slowly with S.

T: We have $6m + 12 = 120$. We always try to get the variables alone on one side of the equals sign and the numbers on the other side of the equals sign. Step 2 shows us how to move the number 12 from the left side to the right side using the additive inverse.

Read step 3 with S. Complete the page. Have S. try the example. Help, if necessary. Stay with this example until S. is completely successful and feels confident with the process. Repeat this procedure for page 31, all steps.

Pages 31–33. Read each example with S. Point out that the secret to simplifying $26 - 2(x - 5) = 72$ is in remembering that $26 - 2$ means $26 + (-2)$. Return to Level 3 Book 1 and review the necessary skills with S. if necessary. As you complete each example, have S. try working it on a separate piece of paper to be sure the skill has been learned before you continue. If it has not been learned, discuss and have S. try it again.

Page 34. Read the directions to Exercise 9 slowly with S. Do the first problem together.

Write: $7(2 + x) = 28$

T: Multiply 7 times 2 and 7 times x.

Write: $14 + 7x = 28$

T: Use the additive inverse to get $0 + 7x$ or $7x$ on the left.

Write:
$$\begin{array}{rcr} 14 + 7x &=& 28 \\ -14 \quad & & -14 \\ \hline 7x &=& 14 \end{array}$$

Find x by dividing both sides by 7, using the multiplicative inverse.

Write: $\frac{7x}{7} = \frac{14}{7}$

$x = 2$

Check by using the 2 for x.

Write: $7(2 + x) = 28$

$7(2 + 2) = 28$

$7(4) = 28$

$28 = 28$ check!

Have S. do problem 2 and check it. If it is correct, have him do the other problems, answering and checking each against the answer key before going on. If S. is still not completely successful, first look together at S.'s work to find the exact cause of difficulty. Determine what steps are not correct. Work together on the area of difficulty. If signs are a problem, return to Level 3 Book 1, Table of Contents to locate the type of sign-combining that is a problem. Review the necessary skills with S. before going on. If solving equations is a problem, return to Chapters 6 and 7 and review those skills before going on.

Activity

Materials: Worksheet 17 and pencil

Procedure:

Do problem 9 together.

Write: $-8c + 4 + 2c = 22$

$-6c + 4 = 22$

T: Add -4 to both sides.

Write:
$$\begin{array}{rcr} -6c + 4 &=& 22 \\ -4 & & -4 \\ \hline -6c \quad &=& 18 \end{array}$$

T: Divide both sides by -6.

Write: $\frac{-6c}{-6} = \frac{18}{-6}$

$c = -3$

T: Check this by substituting -3 for c:

Write: $-8(-3) + 4 + 2(-3) = 22$

$24 + 4 - 6 = 22$

$28 - 6 = 22$

$22 = 22$ check!

If signs are a problem, return to Level 3 Book 1, Table of Contents to locate the type of sign-combining that is a problem. Review the necessary skills with S. before going on to problem 2.

Now have S. do problem 2 and check it. If it is correct, have him do the other problems, answering and checking each against the answer key before going on.

Chapter 10. Solving Equations With Variables on Both Sides

Concepts:

- Knowing to get variables on one side and numbers on the other first
- Knowing how to get variables on one side and numbers on the other
- Knowing how to solve the equation

Pages 35 & 36. Read the discussion slowly with S. For additional explanation of the process, return to the notes in this *Teacher's Guide* for Book 2, Chapter 4. As you complete each example, have S. try working it on a separate piece of paper to be sure the skill has been learned before you continue. If it has not been learned, first look together at S.'s work to find the exact cause of difficulty. Determine what steps are not correct. Then work together on the area of difficulty. If S. is still not completely successful, have him try it again.

Page 37. Read the directions to Exercise 10 with S. Do problem 5 together.

Write: $2(r - 3) = 4(r - 10)$

T: Multiply the items in the parentheses on the left by 2 and the items in the parentheses on the right by 4 (i.e., use the distributive property).

Write: $2r - 6 = 4r - 40$

T: Now use the additive inverse to get the variables on one side and the numbers on the other. Move the r's from the right side to the left by adding $-4r$ to both sides.

Write:
$$\begin{array}{rcl} 2r - 6 &=& 4r - 40 \\ -4r & & -4r \quad \text{(additive inverse)} \\ \hline -2r - 6 &=& -40 \end{array}$$

T: Now move the number -6 to the right by adding $+6$ to both sides.

Write:
$$\begin{array}{rcl} -2r - 6 &=& -40 \\ +6 & & +6 \quad \text{(additive inverse)} \\ \hline -2r &=& -34 \end{array}$$

T: Now divide both sides by -2.

Write: $\frac{-2r}{-2} = \frac{-34}{-2}$ (multiplicative inverse)

$r = 17$

Have S. do problem 1 and check it. If it is correct, have him do the other problems, answering and checking each against the answer key before going on.

Activity

Materials: Worksheet 18 and pencil

Procedure:

Do problem 4 together. Write: $6s - 32 = 2s$

T: Now use the additive inverse to get the variables on one side and the numbers on the other. Move the s's from the left side to the right by adding $-6s$ to both sides.

Write:
$$\begin{array}{rcl} 6s - 32 &=& 2s \\ -6s & & -6s \quad \text{(additive inverse)} \\ \hline -32 &=& -4s \end{array}$$

T: Now divide both sides by -4 to find what s equals.

Write: $\frac{-32}{-4} = \frac{-4s}{-4}$ (multiplicative inverse)

$8 = s$

T: Check this by substituting 8 for each s in the equation.

Write: $6s - 32 = 2s$

$6(8) - 32 = 2(8)$

$48 - 32 = 16$

$16 = 16$ check!

T: When both sides equal each other we know the answer is right. If the check doesn't work, look for sign errors, miscalculations, or distributive errors.

Have S. do problem 1 and check it. If it is correct, have him do the other problems, answering and checking each against the answer key before going on. If S. is still not completely successful, first look together at S.'s work to find the exact cause of difficulty. Determine what steps are not correct. Then work together on the area of difficulty. If signs are a problem, return to Level 3 Book 1, Table of Contents to locate the type of sign-combining that is a problem. Review the necessary skills with S. before going on.

Chapter 11. Solving Literal Equations

Concepts:

- Knowing what a question is asking
- Knowing how to solve for what is asked

Pages 38 & 39. Read the discussion slowly with S. If S. has difficulty, focus instead on the explanation suggested below for Exercise 11.

Page 40. Read the directions to Exercise 11 with S.

T: The first four problems ask us to get x alone. Look at each and see what operation is happening and what we must do to get the x alone. In problem 1 we're subtracting t from the x. To get x alone, we must add t to both sides. In problem 2 we're dividing the x by b. To get x alone, we multiply both sides by b. In problem 3 we're adding y to the x. To get x alone, we add $-y$ to both sides. In problem 4 we're multiplying the x by d. To get x alone, we divide both sides by d.

Have S. do problem 1 and check it. If it is correct, have him do the other problems, answering and checking each against the answer key before going on.

Activity

Materials: Worksheet 19 and pencil

Procedure:

Do the first problem together.

Write: $x + y = z$

T: We're adding y to the x. To get x alone, we add $-y$ to both sides.

Write:

$$\begin{array}{l} x + y = z \\ \underline{\quad - y = z - y} \\ x \quad\;\; = z - y \end{array}$$

Have S. do problem 2 and check it. If it is correct, have him do the other problems, answering and checking each against the answer key before going on. If S. is still not completely successful, first look together at S.'s work to find the exact cause of difficulty. Determine what steps are not correct. Then work together on the area of difficulty.

Post-Test. After you have completed this chapter, administer the post-test. Review the appropriate pages for any incorrect answers. Then retest. When S. has successfully completed the work, go on to the next book.

Introduction

If you have been referred to this book by the diagnostic test, have S. take the pre-test to determine where to begin in this book. On the answer page to the pre-test is a list of the questions and their related chapter pages. When S. has missed a question, turn to the chapter referred to and begin.

If S. is working through the complete series, there is no need to give the pre-test. The diagnostic test has already indicated a lack of skill, and it would be threatening and discouraging to have S. fail again at these questions. Simply begin with Chapter 1.

Chapter 1. Using Formulas to Solve Problems

Concepts:

- Understanding that a formula is a standardized way of showing a relationship among items
- Knowing how to use a formula
- Knowing how to solve a problem using a formula and the given numbers

Using a formula requires S. to substitute given values (numbers) for variables, and then use equation-solving methods to find remaining unknown values. The additive inverse and multiplicative inverse are usually sufficient for such equation solving.

Pages 7 & 8. Read the discussion slowly with S. If S. has difficulty, try this additional explanation.

T: A formula is a set of directions for dealing with a set number of items which always have the same relation to each other. Certain useful formulas include area, circumference, temperature, distance, and interest.

As you complete each example on pages 7 and 8, have S. try working it on a separate piece of paper to be sure the skill has been learned before you continue. If it has not been learned, discuss and have S. try it again. If S. is still not completely successful, first look at S.'s work to find the exact cause of difficulty. Determine what steps are not correct. Then work together on the area of difficulty.

Page 9. Read slowly through Exercise 1 with S. if reading is a problem. Do the first problem together. First, write the formula. Next, substitute the given numbers. Then, solve the equation for the variable we don't yet know.

Write: $d = rt$

$3200 = r(5)$

$\frac{3200}{5} = \frac{r(5)}{5}$

$640 = r$

T: To get r alone, we divide by 5 on both sides. So, the plane is flying at 640 miles per hour (mph).

Have S. do problem 2 and check it. If it is correct, have her do the other problems, answering and checking each against the answer key before going on. If reading is a problem, read slowly with S.

Activity

Materials: Worksheet 20 and pencil

Procedure:

Read the first problem with S.

i = interest (the amount earned on your money)

p = principal (the amount of money you put in)

r = rate (the % paid)

t = time in years (6 months is $\frac{1}{2}$ year; 12 months is 1 year)

Write:

$i = prt$

$i = (500)\left(\frac{8}{100}\right)(1)$

$i = \frac{(500)(8)(1)}{100}$

$i = 40$

This can also be written as follows.

Write: $i = prt$

$i = p \times r \times t$

$i = \overset{5}{\cancel{500}} \times \frac{8}{\underset{1}{\cancel{100}}} \times 1$

So, $i = \$40$

You may need to review percentages with S. If S. is using a calculator with a percent (%) key, write the problem $i = (500)(8\%)(1)$. The percent key automatically divides by 100.

Have S. do problem 2 and check it. If it is correct, have her do the other problems, answering and checking each against the answer key before going on. If solving equations is a problem, return to Level 3 Book 2, Chapters 6 and 7. Review the necessary skills with S. before going on. If reading is a problem, read slowly with S.

Chapter 2. Writing Your Own Equations to Solve Word Problems

Concepts:

- Reading and deciding what you are trying to find out
- Knowing how to write an equation to show the relationships
- Knowing how to solve the equation
- Knowing how to check an answer

Pages 10 & 11. Before reading pages 10 and 11, reassure S. that she doesn't have to memorize this information. It is merely an overview of the process, which will become clearer as she does the problems. Try the following problems with S. before doing pages 10 and 11. Write the problems out and underline key words and numbers as you work.

1. After he lost 13 pounds, Ben weighed 190 pounds. What was his original weight?

   Let w = original weight

   original – 13 = 190

   $w - 13 = 190$

   $w = 190 + 13$

   $w = 203$

2. During a bake sale, Marco made $12 more than Maria. If Marco made $49, how much did Maria make?

   Let c = Maria's income

   $49 = 12 + c$

   $49 - 12 = c$

   $c = 37$

3. A store hired 10 workers this year. This is 4 fewer than they hired last year. How many workers were hired last year?

   Let y = last year's number of workers

   $10 = y - 4$

   $+4 = \quad +4$

   $14 = y$

T: So, last year the store hired 14 workers. Are 10 workers 4 fewer than 14? [S: Yes.]

Let S. stay with these problems until she is clear about:

1. locating the important words and numbers
2. making notes of these important words and numbers
3. "translating" from the words and numbers into an equation
4. solving the equation

Write the above four reminders for S. to study with the preceding problems. Rework these problems until S. is confident. Then, read pages 10 and 11 with S. Have S. work through the problem as you read along together.

Pages 12 & 13. Have S. complete Example 1 and check that S. is following the correct process. If not, look together at S.'s work to find the exact cause of difficulty. Determine what steps are not correct. If reading is a problem, read slowly to S. each time you look for information. Work together on any areas of difficulty and try the example again before continuing on to Example 2. The following notes help explain Examples 1 and 2.

Example 1. If S. doesn't understand that "2 miles is $\frac{1}{4}$ as far as e" should be written algebraically as $2 = \frac{1}{4}e$, then try this.

Write: 2 is $\frac{1}{4}$ "as far as" e

2 is $\frac{1}{4}$ of e

$2 = \left(\frac{1}{4}\right)(e)$

$2 = \frac{e}{4}$

$(4)2 = \frac{e}{4}(4)$

$8 = e$

Note: In Example 2, step 1, the dash is for spacing only. Read such dashes as "which is" ("which is $2b$," etc.).

T: Since Winston collected twice as much as Bath, we can call Bath's amount b and Winston's $2b$. Adams collected three times as much as Bath, so since we called Bath's b, we call Adams's $3b$. Parry collected four times as much as Bath, so since we called Bath's b, we call Parry's $4b$. Let's list these.

Write: Bath's amount = b

Winston's amount = $2b$

Adams's amount = $3b$

Parry's amount = $4b$

T: All together they collected 400 tons. So if we add all these amounts, we will get 400 tons.

Write: $b + 2b + 3b + 4b = 400$

Return to Steps 2, 3, and 4.

Page 14. Read slowly through Exercise 2 with S. if reading is a problem. If S. needs more help, do the following. Reread each problem to find every relevant piece of information.

Problem 1.

Write: Sale price: $60

$60 is 25% off regular price (P)

Note: 25% off is 75% of regular price

$$60 = 75\% \text{ of } P$$
$$60 = \frac{3}{4}(P)$$
$$(4)60 = (4)\frac{3}{4}(P)$$
$$240 = 3P$$
$$\frac{240}{3} = \frac{3P}{3}$$
$$80 = P$$

If S. needs help, look together at the answers on pages 34 and 35 to complete problems 2–6. Now have S. go back and try problem 1 without using the answer key except to check. If it is correct, have her do the other problems, answering and checking each before going on. If S. is still not completely successful, first look together at S.'s work to find the exact cause of difficulty. Determine what steps are not correct. Then work together on the area of difficulty.

Activity

Materials: Worksheet 21 and pencil

Procedure:

For problem 1, write:

First = x

Second = $2x$

Third = $x + 5$

Fourth = $2x + 10$

$(x) + (2x) + (x + 5) + (2x + 10) = 81$

$$6x + 15 = 81$$
$$-15 \quad -15$$
$$6x = 66$$
$$\frac{6x}{6} = \frac{66}{6}$$
$$x = 11$$

Have S. do problem 1 and check it. If it is correct, have S. do the other problems, answering and checking each against the answer key before going on. Have S. use the answers to help work the problems when necessary. Stay with each problem until S. is completely successful. If solving equations is a problem, return to Level 3 Book 2, Chapters 6 and 7 and review the necessary skills with S. before going on.

Chapter 3. Using a Graphic Organizer to Solve Word Problems

Concepts:

- Understanding how to translate word problems into math equations
- Recognizing math in everyday life
- Understanding how to use a graphic organizer to help understand and solve word problems

Pages 15 & 16. Read the discussion and ask if S. has any questions. Explain the graphic organizer, and tell S. that it is just another tool to help S. translate a word problem into an equation by identifying what S. needs to do. Walk S. through how it relates to the 6 questions.

1. Identify the question.
2. Label the necessary and unnecessary information.
3. Write an equation.
4. Solve the problem.
5. Check your answer.
6. Make sure your answer makes sense.

Draw on the board the graphic organizer from page 11 in the Introduction to this guide. Use it to help answer the example problem. You may copy page 11 for S. Have S. use a copy of the graphic organizer to complete Exercise 3.

Chapter 4. Solving Multiple-Choice Set-Up Problems

Concepts:

- Knowing how to answer multiple-choice questions
- Distinguishing the correct answer from wrong answer choices
- Understanding the use of set-up problems

Pages 17 & 18. Read the discussion. Ask if S. has any questions or concerns. Read the example and write it on the board. Discuss the use of different variables instead of x, and how an equation can be equal even if the parts are not in the same order. Do the example; write the whole question on the board including the answer choices. Discuss how the variables could be different, that is, instead of j, it could be x or m or w or whatever you want it to be, and the same with b. Now have S. figure out other ways to write $j - 5 = b$ that are equivalent.

$x - 5 = y$, if Jess = x and Bob = y

$w - 5 = z$, if Jess = w and Bob = z

$b = j - 5$, if Jess = j and Bob = b

$j = b + 5$, if Jess = j and Bob = b

Make sure that S. understands that he cannot change the meaning of an equation.

Page 19. Read the instructions to Exercise 4. Have S. answer all questions by filling in the bubble for the answer.

Explain that wrong answers are often developed by following the correct answer steps but applying common errors that people often commit. So, S. needs to be careful in doing her computations.

Chapter 5. Ratio

Concepts:

- Understanding that a ratio is a comparison of two numbers
- Knowing how to write a ratio as a fraction or with a colon (:)
- Knowing how to reduce ratios

Note: The dash on page 20 is for spacing only. Read "$\frac{3}{1}$" and "3:1" as "3 to 1". Also, in advanced math, ratios can include more than two numbers. For the purposes of this series, a ratio is defined as including two numbers only.

Pages 20–22. Read the discussion slowly with S. Point out that the biggest difference between a ratio and a rate is that a rate is reduced until one of the terms is the number one. For example, miles per gallon means number of miles to one gallon. Remind S. that the order of terms in a ratio is important. Writing verbal ratios first can help to avoid errors. For Example 1, pages 20 and 21, write:

$$\frac{\text{won}}{\text{played}} = \frac{96}{144} = \frac{2(48)}{3(48)} = \frac{2}{3}$$

Note that ratios are always written in reduced form.

Point out also that the bar used for ratios has three separate uses:

1. fractions (the number of parts/total number of parts)
2. division (dividend/divisor)
3. ratio (comparison of numbers: first number/second number)

A ratio is like a fraction only to a certain point. An improper fraction can be written as a mixed number or whole number, unlike a ratio. For example, $\frac{5}{3}$ equals $1\frac{2}{3}$ as a fraction, but $\frac{5}{3}$ equals only $\frac{5}{3}$ as a ratio.

As you complete each example, have S. try working it on a separate piece of paper to be sure the skill has been learned before you continue. If it has not been learned, discuss and have S. try it again. If reducing fractions is a problem, return to Level 2 Book 1, Chapter 8 and review the necessary skills before you continue.

Page 23. Read slowly through Exercise 5 with S. if reading is a problem. In problem 4, remind S. that 1 foot = 12 inches, so $1\frac{1}{2}$ feet = 18 inches.

Have S. do problem 1 and check it.

Write: 1 foot = 12 inches

$\frac{1}{2}$ foot = 6 inches

$1\frac{1}{2}$ feet = 18 inches

In inches:

$\frac{18}{6} = \frac{3}{1}$

If it is correct, have S. do the other problems, answering and checking each against the answer key before going on. Have S. use the answer key on page 36 to help work the problems when necessary.

Note: Problems 1–4 may also be answered using a colon as follows.

1a. 56:84 = 2:3

1b. 28:84 = 1:3

2a. 15:20 = 3:4

2b. 5:20 = 1:4

2c. 15:5 = 3:1

3. 20:10 = 2:1

4. 18:6 = 3:1 or $1\frac{1}{2}:\frac{1}{2}$ = 3:1

Activity

Materials: Worksheet 22 and pencil

Procedure:

Have S. do problem 1 and check it. If it is correct, have her do all the problems, answering and checking each against the answer key. Point out that from now on exercises will include multiple-choice questions. If S is still not completely successful, first look together at S.'s work to find the exact cause of difficulty. Determine what steps are not correct. Then work together on the area of difficulty.

Chapter 6. Proportion

Concepts:

- Understanding that a proportion is an equation made up of two ratios (that one ratio equals another ratio)
- Knowing to cross multiply the fractions or outer and inner terms (so that the products are equal)
- Knowing how to solve algebraic ratios

Pages 24–27. Read the discussion slowly with S., stressing the chapter concepts. Emphasize to S. that a true proportion is an equation with two ratios.

Note: The dash in the fifth line on page 24 is for spacing only.

You may prefer to work only with fractions on pages 24–27. The colon may confuse some students.

To help explain Example 1 on page 25, have S. write a word ratio first to keep terms aligned:

$\frac{\text{miles}}{\text{hour}} = \frac{200}{5} = \frac{x}{8}$

(known ratio) (ratio with one part unknown)

Note: You may wish to complete the following activity, Worksheet 23, with whole numbers before progressing to Examples 2 and 3 (pages 26 and 27) that include fractions.

Activity

Materials: Worksheet 23 and pencil

Procedure:

Read each problem with S. Since there are several things to do, call her attention to them. Remind her of them if she forgets. If S. is not doing well, return to the chapter in the book and review it. If S. wishes, copy this page for her to use for practice at home.

If S has difficulty, try this additional explanation.

T: Here are two ways to tell if fractions are equivalent. You can reduce both to the same fraction:

$\frac{2}{4} = \frac{3}{6}$

$\frac{1}{2} = \frac{1}{2}$

Or you can use cross-products as a shortcut:

12 $\frac{2}{4} = \frac{3}{6}$ 12

Since 12 = 12, the fractions are equivalent. You can use either of these methods to help find a missing part of a proportion.

Pages 26 & 27. As you complete each example on pages 26–27, have S. try working it on a separate piece of paper to be sure the skill has been learned before you continue. If S. is still not completely successful, look together at S.'s work to find the exact cause of difficulty. Determine what steps are not correct. Then work together on the area of difficulty. Discuss and have S. try it again. If solving equations is a problem, return to Level 3 Book 2, Chapters 6 and 7 and review the necessary skills with S. before going on.

Page 28. Read slowly through Exercise 6 with S. if reading is a problem. Have S. do problem 1 and check it. If it is correct, have S. do the other problems, answering and checking each against the answer key before going on. If S. is still not completely successful, look together at S.'s work to find the exact cause of difficulty. Determine what steps are not correct. Then work together on the area of difficulty. Discuss and have S. try it again. If solving equations is a problem, return to Level 3 Book 2, Chapters 6 and 7 and review the necessary skills with S. before going on.

Chapter 7. Proportion Problems With Added Steps

Concepts:

- Knowing that the measurements must agree in a ratio
- Knowing that sometimes units must be converted or several unknowns identified before setting up a ratio

Pages 29 & 30. Read the discussion with S., stressing the chapter concepts. If S. has difficulty, try working Example 1 on page 29 as follows.

Equation: Sam + Jo

(hr.) ($/hr) + (hr.) ($/hr.) = total $

Let x = hourly wage

$3x$ = Sam's pay

$2x$ = Jo's pay

$3x + 2x = 120$

$5x = 120$

$x = 24$

Sam's pay = $72; Jo's pay = $48

Using a variable to identify the missing part is helpful. This type of problem is commonly written in terms of ratio. If the ratio of Sam's pay to Jo's pay is 3:2, then Sam's actual pay is $3x$ and Jo's is $2x$. Refer to the notes for Chapter 5 on page 33 of this *Teacher's Guide* to review ratio if necessary.

Page 31. Read slowly through Exercise 7 with S., if reading is a problem. Have S. do problem 1 and check it. If it is correct, have S. do the problems, answering and checking each against the answer key before going on. If S. is still not completely successful, first look together at S.'s work to find the exact cause of difficulty. Determine what steps are not correct. Then work together on the area of difficulty. If solving equations is a problem, return to Level 3 Book 2, Chapters 6 and 7 and review the necessary skills with S. before going on.

Activity

Materials: Worksheet 24 and pencil

Procedure:

Read each problem with S. Since there are several things to do, call her attention to them. Remind her of them if she forgets. If S. is not doing well, return to the chapter in the book and review it. If S. wishes, copy this page for her to use for practice at home.

Post-Test. After you have completed this chapter, administer the post-test. Review the appropriate pages for any incorrect answers. Then retest. When S. has successfully completed the work, go on to the next book.

Level 3 Book 4: Exponents, Roots, and Polynomials

Introduction

If you have been referred to this book by the diagnostic test, have S. take the pre-test to determine where to begin in this book. On the answer page to the pre-test is a list of the questions and their related chapter pages. When S. has missed a question, turn to the chapter referred to and begin.

If S. is working through the complete series, there is no need to give the pre-test. The diagnostic test has already indicated a lack of skill, and it would be threatening and discouraging to have S. fail again at these questions. Simply begin with Chapter 1.

Chapter 1. Factors

Concepts:

- Knowing that a *product* is the result of multiplication
- Knowing that each of the items that are multiplied is a *factor*
- Knowing that a product may be made by many factor combinations
- Knowing that factors may be positive or negative
- Recognizing ± as *positive or negative*

Pages 7 & 8. Read the discussion slowly with S., stressing the chapter concepts. Remind S. that parentheses ( ) tell him to multiply. Point out that we use pairs of factors unless we are looking for prime factors. For example:

–16

–1, 16
1, –16
–2, 8
2, –8
4, –4

Page 8. Read the directions to Exercise 1 with S. Remind S. that he can multiply in any order and the product will be the same. For example, (2)(–4) = (–2)(4) = –(2)(4), so it may be written any of these ways. In order to obtain a negative product, the factors must have opposite or different signs. So, in problems 3 and 5, if the first number is negative, the second must be positive (and vice versa):

(–1)(20)
(1)(–20)
(–2)(10)
(2)(–10)
(–4)(5)
(4)(–5)

Have S. do problem 1 and check it. If it is correct, have S. do the other problems, answering and checking each against the answer key before going on. Explain to S. that answers to problems 3 and 5 could also be written as follows:

1, –25
–1, 25
–5, 5

If multiplying is a problem, return to Level 1 Book 4 and review the necessary skills with S. before going on.

Activity

Materials: Worksheet 25 and pencil

Procedure:

Do problem 1 with S. If he needs help, look together at the answer key. Then have S. go back and try problem 1 without using the answer key except to check. If it is correct, have him do the other problems, answering and checking each before going on.

Chapter 2. Exponents

Concepts:

- Knowing what a base number and exponent (power) are
- Knowing how the exponent affects the base numbers

Pages 9–11. Read the discussion slowly with S., stressing the chapter concepts. Point out that exponents only act on what is immediately in front of them, e.g., $5^3 = (5)(5)(5)$; $(xy)^2 = (xy)(xy)$. Note also that a base is any number that is to be multiplied by itself, e.g., 5^2: the base is 5; $(xy)^2$: the base is (xy). Mention to S. that there are several ways of saying "power." 5^2 can be read:

5 *squared*

5 *to the power of 2*

5 *to the second power*

5 *to the second*

Page 11. Read the directions to Exercise 2 with S. Have S. do problem 1 and check it. If it is correct, have S. do the other problems, answering and checking each against the answer key before going on. If multiplying is a problem, return to Level 1 Book 4. If signs are a problem, return to Level 3 Book 1, Table of Contents to locate the type of sign-combining that is a problem. Review the necessary skills with S. before going on. Have S. copy the problems and answers to do for practice at home.

Activity

Materials: Worksheet 26 and pencil

Procedure:

Read the directions slowly with S., pointing out each direction at the top of each column. Work the problems together. Check each answer to be sure the skill has been learned before you continue to the next problem. If multiplying is a problem, return to Level 1 Book 4 and review the necessary skills with S. before going on.

Chapter 3. Square Roots

Concepts:

- Recognizing the term *square root* and the symbol $\sqrt{\ }$
- Knowing that a number times itself equals a square
- Knowing some common squares
- Knowing how to use a square-root chart

Pages 12 & 13. Read the discussion on page 12 slowly with S. Look together for at least six square roots (randomly) on page 13. Reassure S. that these do not have to be memorized. The more common (whole number) ones will be learned in the activity that follows.

Remind S. that the square root of a number is a factor of that number. To find the square root of 36, ask, "What factor of 36, times itself, is 36?"

Activity

Materials: Worksheet 27, pencil, and calculator if possible

Procedure:

Have S. do problem 1 and check it. If it is correct, have him do the other problems, answering and checking each against the answer key before going on. Remind S. to do the work in parentheses before squaring any numbers in problem 2.

Calculators, even fairly inexpensive ones, do square roots. They generally have a key with $\sqrt{\ }$ on it. To find a square root using a calculator, press the number and then the $\sqrt{\ }$ key. The calculator will tell you the square root of the number. It won't be rounded off, however.

Try Exercise 3 on page 13 using a calculator. For problem 1, press 12. Press $\sqrt{\ }$. You get 3.4641016.

Note that it's not rounded off. Have S. do the others the same way (except problem 4). Check them against the answer key.

Page 13. This time, do Exercise 3 as indicated, without using a calculator. Read the directions with S. Have S. do problem 1 and check it. If it is correct, have S. do the other problems, answering and checking each against the answer key before going on.

Chapter 4. Using a Calculator to Find Square Roots and Exponents

Concepts:

- Understanding how to use a calculator to find square roots and exponents

Pages 14 & 15. Have S. take out her calculator and have her look at it as you read the discussion. The calculator used is the TI-30XS, which is the calculator used when taking the GED. Point out where the keys are and what they do. Limit the discussion to just the 5 function keys that are needed. Do the examples in the chart with S., watching to make sure she presses the correct keys. Have S. do the four examples and talk through the answers while using the calculator. This is an example of the Think-Aloud technique described in the Introduction on page 9.

Page 15. Have S. do Exercise 4. Check answers.

Chapter 5. Terms

Concepts:

- Knowing what terms are
- Knowing that parts of a term are held together by multiplying or dividing
- Knowing that several terms are held together by adding or subtracting
- Knowing that a monomial is a single term
- Knowing that a polynomial is made up of more than one term
- Knowing what *like* and *unlike* terms are

Pages 16 & 17. Read the discussion slowly with S., stressing the chapter concepts. Remind S. that it is the parts of a term that are held together by multiplication or division. As you do the examples, circle the terms (including the sign in front of each term). Point out that like terms have the same base and the same exponent, e.g., $2x^3$ and $7x^3$ are like terms, while $2x^3$ and $7x$ are unlike terms.

Page 17. Read the directions to Exercise 5 with S. Have S. do problem 1, circling each term first. If it is correct, have S. do the other problems, answering and checking each against the answer key before going on.

Activity

Materials: Worksheet 28 and pencil

Procedure:

Read the directions slowly with S. Have S. try each problem, checking it as it is completed. Remind S. as he circles terms that he may include the sign in front of each term.

Chapter 6. Adding and Subtracting Monomials

Concepts:

- Knowing that a monomial is a single term
- Knowing how to combine (add or subtract) like terms
- Knowing how to combine like terms by adding or subtracting the numbers and using the same variables in the answer

Pages 18–20. Read the discussion slowly with S., stressing the chapter concepts. As you complete each example, have S. try working it on a separate piece of paper to be sure the skill has been learned before you continue. If it has not been learned, discuss and have S. try it again. If necessary, return to the notes in this *Teacher's Guide* for Level 3 Book 1 about adding and subtracting signed numbers and review the skills. If S. still has difficulty, try this additional explanation.

T: You want to make some fruit baskets. You have apples, oranges, and pears. In the first basket, you put 5 apples, 4 oranges, 3 apples, and a pear. So there are 8 apples, 4 oranges, and 1 pear in the basket. The 5 apples and 3 apples are added into a single number because they are alike. You can't just say 13 "appleorangepears" because there are three kinds of fruit.

In algebra, like terms are similar to the apples: $5xy^2$ and $3xy^2$ can be combined into a single term, $8xy^2$.

Note: You may wish to complete the following activity before progressing to Exercise 6 on page 20.

Activity

Materials: Worksheet 29 and pencil

Procedure:

Read the directions slowly with S. Have S. try each problem, checking it as it is completed. In problems 9–18, it may help S. to mark all like terms using different colored pens or by coding each like term, for example:

$$\textcircled{+4x^2y} \underline{+ 3xy} \underline{\underline{\underline{+ 5x^2}}} \underline{- 2xy} \textcircled{- 6x^2y} \underline{\underline{\underline{+ 3x^2}}}$$

This makes it easier to see all like terms and harder to omit any. If signs are a problem, return to Level 3 Book 1, Table of Contents to locate the type of sign-combining that is a problem. Review the necessary skills with S. before going on.

Page 20. Read the directions to Exercise 6 with S. Have S. do problem 1 and check it. Remind S. first to mark all like terms. If problem 1 is correct, have S. do the other problems, answering and checking each against the answer key before going on. For problem 7, remind S. that if the denominators of fractions are the same, only the numerators are combined. The denominators remain the same. If signs are a problem, return to Level 3 Book 1, Table of Contents to locate the type of sign-combining that is a problem. Review the necessary skills with S. before going on. Have S. copy the problems and answers to do for practice at home.

Chapter 7. Working Backwards to Solve Multiple-Choice Questions

Concepts:

- Understanding how to answer multiple-choice questions
- Understanding that working backwards is a way to help determine the answer if you are stuck

Pages 21 & 22. Read the discussion. Ask S. for any questions or concerns he has about the instruction. Read the example and write it on the board. Do the math step-by-step and discuss answer choices. Ask S. if he can tell why the wrong answers are wrong.

Page 23. Read the instructions to Exercise 7. Have S. answer all questions by filling in the bubble for the correct answer.

Chapter 8. Multiplying Monomials

Concepts:

- Knowing that to multiply a monomial by another monomial you begin by multiplying the numbers (with their signs) and then the variables (alphabetically)
- Knowing that x is x^1
- Knowing to add exponents in multiplying variables

Pages 24–27. Read the discussion slowly with S., stressing the chapter concepts and those processes with arrows and printed in bold. If S. has difficulty seeing bold, use a color in its place. Be sure that S. does not add exponents when adding like terms. Also, it may be easier for S. if you ask him to find the product's sign first, then its numerical coefficient, and its variables alphabetically, with exponents.

If S. has difficulty following the discussion, try this additional explanation.

T: Combining like terms was compared to putting fruit in baskets. Multiplying and dividing monomials is like making fruit salad. The various pieces of fruit are cut up (factored) and the parts rearranged.

Remember that x is really $1x^1$.

To multiply monomials, look first for like factors and multiply them. Then multiply all the unlike products to get the final product. When there are many exponents in use, it helps to write each monomial first without the exponents, and then rearrange the factors so multiplying like factors is easier. For example: $(4x^3y^4)(-2x^2y^3) = (4xxxyyyy)(-2xxyyy) = (4)(-2)(xxxxx)(yyyyyyy) = -8x^5y^7$. Once you know how many x's and y's there are to be multiplied, you can use that number for the exponent ($xxxxx = x^5$).

Note: You may wish to complete the following activity before progressing to Exercise 8 on page 27.

Activity

Materials: Worksheet 30 and pencil

Procedure:

Read the directions slowly with S. Have S. try each problem, checking it against the answer key as it is completed. If signs are a problem, return to Level 3 Book 1, Table of Contents to locate the type of sign work that is a problem. Review the necessary skills with S. before going on.

Page 27. Read the directions to Exercise 8 with S. Have S. do problem 1 and check it. If it is correct, have him do the other problems, answering and checking each against the answer key before going on. Point out that from now on, exercises will include multiple-choice problems. If signs are a problem, return to Level 3 Book 1, Table of Contents to locate the type of sign work that is a problem. Review the necessary skills with S. before going on. Have S. copy the problems and answers to do for practice at home.

Chapter 9. Dividing Monomials

Concepts:

- Knowing that to divide a monomial by another monomial you begin by dividing the numbers (with their signs) and then the variables (alphabetically)
- Knowing that x is x^1
- Knowing to subtract exponents in dividing variables

Pages 28–30. Read the discussion slowly with S., stressing the chapter concepts. If S. has difficulty, try these additional directions for division.

1. Write both the top and bottom monomials as factors without using exponents.
2. Find common factors above and below the division bar.
3. Divide each of these factor pairs.
4. Multiply the "leftover" factors above and below the division bar.

This is the simplest form of the quotient (result of division).

For example:

$$\frac{a^4b^2c^3}{a^2bc}$$

1. $\frac{aaaabbccc}{aabc}$
2. $\frac{a}{a}\ \frac{a}{a}aa\ \frac{b}{b}b\ \frac{c}{c}cc$
3. $\frac{a}{a} = 1 \quad \frac{a}{a} = 1 \quad \frac{b}{b} = 1 \quad \frac{c}{c} = 1$
4. $\frac{(1)(1)aa(1)b(1)cc}{(1)(1)\ \ (1)\ (1)} = \frac{a^2bc^2}{1}$ or a^2bc^2

S. may find it simpler to think of the process as dividing common factors to get a quotient of 1 than as a process of cancelling. Remind S. that any number divided by itself equals the number one. Be sure also that S. remembers the sign rules for division. If the sign rules are a problem, return to Level 3 Book 1 and review the necessary skills with S.

Page 31. Read the directions to Exercise 9 with S. Have S. do problem 2 and check it. If it is correct, have S do the other problems, answering and checking each before going on. If signs are a problem, return to Level 3 Book 1, Table of Contents to locate the type of sign work that is a problem. Review the necessary skills with S. before going on. Have S. copy the problems and answers to do for practice at home.

Activity

Materials: Worksheet 31 and pencil

Procedure:

Read the directions slowly with S. Have S. try each problem, checking each as it is completed. If signs are a problem, return to Level 3 Book 1, Table of Contents to locate the type of sign work that is a problem. Review the necessary skills with S. before going on.

Chapter 10. Adding Polynomials Together

Concepts:

- Knowing that polynomials have more than one term
- Knowing to add like terms only
- Knowing how to set up columns of like terms

Pages 32 & 33. Read the discussion slowly with S., stressing the chapter concepts. Point out to S. that we write the terms of the sum with the alphabetically highest exponent first, followed by terms in descending order of exponents. Look at this order as shown in Examples 1 and 2.

For example: $4x^2 - 3xy + 6z^2$

T: Note that the parentheses are used here to show that groups of terms were originally part of different polynomials. Each polynomial represents a single, unique number value.

When adding numbers, you don't make any changes in their signs, so you don't really need to keep the polynomials separated. You can erase the parentheses and combine like terms to find the simplest form of the sum. For example:

$(3c^2 - 4c - 7) + (2c^2 + 2c + 2) =$

$3c^2 - 4c - 7 + 2c^2 + 2c + 2 = 5c^2 - 2c - 5$

Page 33. Read the directions to Exercise 10 with S. Have S. do problem 2 and check it. If it is correct, have S. do the other problems, answering and checking each against the answer key before going on. If knowing what like terms are is a problem, review pages 16 and 17 before going on. If signs are a problem, return to Level 3 Book 1, Table of Contents to locate the type of sign-combining that is a problem. Review the necessary skills with S. before going on. Have S. copy the problems and answers to do for practice at home.

Activity

Materials: Worksheet 32 and pencil

Procedure:

Read the directions slowly with S. Have S. try each problem, checking it as it is completed. If signs are a problem, return to Level 3 Book 1, Table of Contents to locate the type of sign-combining that is a problem. Review the necessary skills with S before going on.

Chapter 11. Subtracting Polynomials

Concepts:

- Knowing that the rules for subtracting polynomials are the same as for subtracting signed numbers: add the opposite of each term in the second polynomial
- Knowing to subtract like terms only
- Knowing how to set up columns of like terms

Pages 34 & 35. Read the discussion slowly with S., stressing the chapter concepts, especially on page 34, Step 2. Remind S. that each polynomial represents a unique number value. When you subtract signed numbers and like terms, you add the opposite of the second number or term (the amount being taken away). Since the second polynomial represents a single number, you have to add the opposite of each term in the polynomial. For example:

$(4x + 3y) - (2x - 4y) =$

$(4x + 3y) + (-2x + 4y) =$ ← add opposite

$4x + 3y - 2x + 4y =$ ← combine like terms

$2x + 7y$

If your sum has terms with the same bases and different exponents, write the sum with the largest exponents first.

Page 35. Read the directions to Exercise 11 with S. Have S. do problem 1 and check it. If it is correct, have S. do the other problems, answering and checking each against the answer key before going on. If signs are a problem, return to Level 3 Book 1, Table of Contents to locate the type of sign-combining that is a problem. Review the necessary skills with S. before going on. If S. is still not completely successful, first look together at S.'s work to find the exact cause of difficulty. Determine what steps are not correct. Then work together on the area of difficulty. Have S. copy the problems and answers to do for practice at home.

Activity

Materials: Worksheet 33 and pencil

Procedure:

Remind S. that for subtracting, after lining up like terms, we must remember to add the opposite of all the terms in the bottom line. If combining like terms is a problem, return to Level 3 Book 1 and review the skill with S. before going on.

Chapter 12. Multiplying Polynomials by Monomials

Concepts:

- Knowing that you multiply each term of the polynomial by the monomial
- Knowing how to multiply monomials
- Knowing how to multiply signed numbers

Remind S. of the use of the distributive property in Level 3 Book 2, Chapters 9 and 10. Review the notes for those chapters in this *Teacher's Guide* if necessary.

Pages 36 & 37. Now read the discussion slowly with S., stressing the chapter concepts and those processes with arrows and printed in bold. It is important to remember that a polynomial represents a single number, as does a monomial. You multiply entire numbers by entire numbers, not parts. This is why the distributive property is used.

Page 38. Read the directions to Exercise 12 with S. Have S. do problem 1 and check it. If it is correct, have S. do the other problems, answering and checking each against the answer key before going on. If signs are a problem, return to Level 3 Book 1, Table of Contents to locate the type of sign work that is a problem. Review the necessary skills with S. before going on. If S. is still not completely successful, first look together at S.'s work to find the exact cause of difficulty. Determine what steps are not correct. Then work together on the area of difficulty. Have S. copy the problems and answers to do for practice at home.

Activity

Materials: Worksheet 34 and pencil

Procedure:

Read the directions with S. Have S. try each problem, checking it as it is completed. If multiplying is a problem, return to Level 1 Book 4. If signs are a problem, return to Level 3 Book 1, Table of Contents to locate the type of sign work that is a problem. Review the necessary skills with S. before going on.

Chapter 13. Multiplying Polynomials by Polynomials

Concepts:

- Knowing to multiply each term of one polynomial by each term of the other polynomial
- Knowing that these may be multiplied vertically or horizontally
- Knowing to combine like terms
- Knowing when and how to combine like terms

Pages 39–41. Remind S. that when multiplying polynomials by polynomials, all parts (terms) of each number (polynomial) must be multiplied. Keep the polynomial inside parentheses as a reminder that it is a single number or factor. Read the discussion slowly with S., stressing the chapter concepts and those processes with arrows and printed in bold. Stress that we must pay careful attention to the signs, multiplying them and putting the correct sign between each term in the answer. Try showing the problem on page 39 this way:

If we were multiplying two two-digit numbers such as 11 × 34, we would first multiply the top numbers by 4, and then by 3. Then we would add to get the final product.

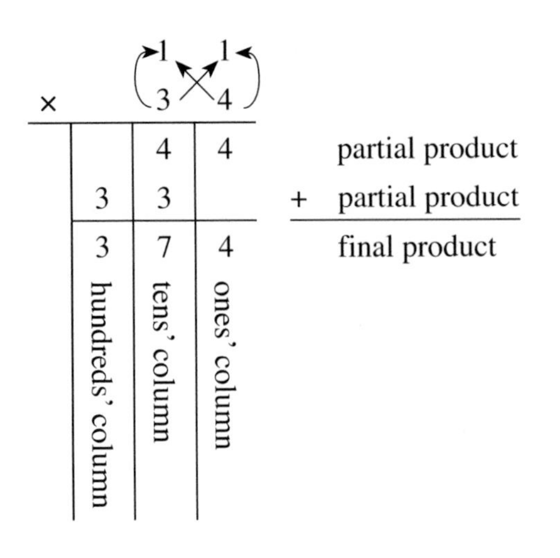

Here's what it looks like when you use the same process to multiply terms of polynomials:

	x	$+2$	
	x	-1	
	$-1x$	-2	partial product
x^2	$+2x$		+ partial product
x^2	$+x$	-2	final product

Use this model to help S. complete Examples 1 and 2. Stay with the examples until S. is completely successful and feels confident.

Page 41. Read the directions to Exercise 13 with S. Have S. do problem 1 and check it. If it is correct, have S. do the other problems, answering and checking each against the answer key before going on. If S. is still not completely successful, first look together at S.'s work to find the exact cause of difficulty. Determine what steps are not correct. Then work together on the area of difficulty. Have S. copy the problems and answers to do for practice at home.

Activity

Materials: Worksheet 35 and pencil

Procedure:

Read the directions slowly with S. Have S. try each problem, checking it as it is completed. Stress that we must be very careful to pay attention to the signs, multiplying the terms and putting the correct sign between each term in the answer. If S. has difficulty, return to Exercise 13 in Book 4. It is usually necessary to practice several of these frequently and repeatedly in order to retain this very complex skill. Have S. copy the problems and answers to do for practice at home.

Chapter 14. Dividing Polynomials by Monomials

Concepts:

- Knowing to divide one term at a time
- Knowing that division of exponents is done by subtracting the exponents and dividing the numbers
- Knowing to divide fractionally with expanding and canceling

Pages 42 & 43. Read the discussion slowly with S., stressing the chapter concepts and the terms printed in bold. If S. has difficulty, try this additional explanation.

T: In chapter 9, when we divided monomials by monomials, we wrote each monomial in factored form and found matching pairs of factors to divide by each other. The leftover factors were the quotient. You can use the same idea when dividing polynomials. First, write all polynomials in factored form. Then find matching pairs of factors to divide in the polynomial and the monomial. The leftover factors are the quotient. Remember to keep the leftover polynomial factor in parentheses as a reminder that it represents a single factor. For example:

$\frac{3a + 3b}{3} = \frac{3(a + b)}{3}$ ⟵ Write in factored form.

$= \frac{\overset{1}{\cancel{3}}(a + b)}{\underset{1}{\cancel{3}}}$ ⟵ Divide common factors.

$= (a + b)$ ⟵ Answer

Page 43. Read the directions to Exercise 14 with S. Have S. do problem 1 and check it. If it is correct, have S. do the other problems, answering and checking each before going on. If S. is still not completely successful, first look together at S.'s work to find the exact cause of difficulty. Determine what steps are not correct. Then work together on the area of difficulty. If S. is still having difficulty or just wants to see another method, do Exercise 14 this way:

Write: $\frac{10m^3 - 15mn}{5m} = \frac{10m^3}{5m} - \frac{15mn}{5m}$

$= \frac{10mmm}{5m} - \frac{15mn}{5m}$

$= 2m^2 - 3n$

Have S. try problem 1 this way. Stay with this problem until S. is completely successful and feels confident. Then have him go on. Have S. copy the problems and answers to do for practice at home.

Activity

Material: Worksheet 36 and pencil

Procedure:

Read the directions slowly with S. Have S. try each problem, checking it against the answer key as it is completed. Have S. copy the problems and answers to do for practice at home.

Chapter 15. Finding Factors of Terms With Variables

Concepts:

- Knowing that parentheses indicate multiplication
- Knowing what *factors* and the *highest common factor* are
- Knowing how to find the highest common factor
- Knowing how to divide by the highest common factor to get the leftover polynomials

Pages 44–46. Read the discussion slowly with S., stressing the chapter concepts. Point out to S. that he can factor the number just as he did in Chapter 1 of Level 3 Book 4, pages 7 and 8. Review those pages with S. if necessary. Have S. try each example on pages 44 and 45. Point out that by multiplying the factors of the answer together, we should get the original polynomial and that this checking method is very useful. Have S. try checking Examples 1 and 2 on page 45 before going on.

If S. still has difficulty, try this additional explanation.

T: The first thing to do when factoring a polynomial is to look for common factors in all of its terms. In order to be a common factor, the signed number or variable must be a factor of each and every term in the polynomial. For example:

Write: $2a - 4b$

T: We can rewrite this polynomial as follows.

Write: $(2)(a) - (2)(2)(b)$

T: This shows all the factors of each term. Are there any factors that are present in both $2a$ and $4b$? [S: 2] Good. Now, when we write $2a - 4b$ in factored form, we write the common factor, 2, outside the parentheses and the leftover term inside the parentheses.

Write: $2(a - 2b)$

T: Remember that 2( ) means multiply everything inside the parentheses by 2. Were there any common variable factors in $2a - 4b$? [S: No.]

T: Here's another example.

Write: $x^2 + 3x$

$(x)(x) + (3)(x)$ ← All factors of terms

← Find common factor

$x(x + 3)$ ← Write common factor outside parentheses and leftover terms inside

T: To check, multiply x by $(x + 3)$. Use the distributive property to see if we get the original polynomial.

Write: $x(x + 3) = x^2 + 3x$

T: So, our factoring was correct.

Page 46. Read the directions to Exercise 15 with S. Have S. do problem 1 and check it. If it is correct, have S. do the other problems, answering and checking each against the answer key before going on. If S. is still not completely successful, first look together at S.'s work to find the exact cause of difficulty. Determine what steps are not correct. Then work together on the area of difficulty.

Checking helps especially when one of the factors must be +1 or −1, as on page 46, problems 3 and 6. Have S. copy the problems and answers to do for practice at home.

Activity

Materials: Worksheet 37 and pencil

Procedure:

Point out to S. that he can factor the number just as he did in Chapter 1 of Level 3 Book 4. Review those pages with S. if necessary. Point out as above that by multiplying the factors of the answer together, we should get the original polynomial and that this checking method is very useful. If signs are a problem, return to Level 3 Book 1, Table of Contents to locate the type of sign work that is a problem. Review the necessary skills with S. before going on.

Chapter 16. Multiplying the Sum and Difference of Two Numbers

Concepts:

- Knowing what a *binomial* is
- Knowing how to multiply two binomials vertically or horizontally
- Knowing that only when multiplying the sum and the difference of two terms will the middle terms combine to equal zero, and thus be eliminated
- Knowing the shortcut for multiplying the sum and the difference of two terms

Pages 47 & 48. Point out to S. that a binomial is a polynomial with exactly two terms in it. (*poly* means *many,* as in *polytechnic* or *polygamy. Bi* means *two,* as in *bicycle* or *biweekly*.) Read the discussion slowly with S. Turn back to Chapter 13 on page 39 and review the vertical and horizontal processes for multiplying polynomials.

Read the directions to Exercise 16 with S. Have S. try multiplying the polynomials vertically and horizontally so that the elimination of the middle terms becomes obvious. Start with problem 2. Then go back and look at problem 1 together.

T: When you multiply $(m + n^2)(m - n^2)$ together, you get $m^2 - n^4$. Notice that both binomials (two-term polynomials) begin with m and both end with n^2. If we square m, we get m^2, and if we square n^2, we get n^4. Then, by putting the minus sign between them, we have our answer $m^2 - n^4$ using the shortcut.

Remind S. to square each part of the term: $(\text{first term})^2 - (\text{second term})^2$. If squaring numbers is a problem, return to Level 3 Book 4, Chapter 2 and review the skill with S.

Activity

Materials: Worksheet 38 and pencil

Procedure:

Have S. try the first three problems vertically, then horizontally, and lastly using the shortcut. If S. has mastered the shortcut, he may want to work the other problems that way only. If S. has difficulty, return to Level 3 Book 4, pages 47 and 48 and work each example together. Then redo Exercise 16 on page 48 of Book 4 using each of the three methods for each problem. Finally, return to Worksheet 38 and redo it.

Chapter 17. Factoring the Difference of Two Squares

Concepts:

- Knowing squares of numbers and variables
- Recognizing squares
- Recognizing when a binomial is the difference of two squares
- Knowing how to "undo" the difference of two squares

Pages 49 & 50. Read the discussion slowly with S., stressing the chapter concepts. If recognizing "perfect squares" is a problem, return to the activity for Worksheet 27 on page 36 of this *Teacher's Guide* and review.

Page 51. Read the directions to Exercise 17 with S. Have S. do problem 1 and check it. If S. has difficulty, return to Level 3 Book 4, pages 47 and 48 and work the examples and Exercise 14 together, using each of the three methods for each problem. Remind S. that factoring is "undoing" multiplying.

If recognizing perfect squares is a problem, return to the activity for Worksheet 27 on Page 36 of this *Teacher's Guide* and review. Then try page 51 again, checking each problem before going on.

Activity

Materials: Worksheet 39 and pencil

Procedure:

Read the directions with S. Have S. do problem 1 and check it. If S. has difficulty, return to Level 3 Book 4, pages 47 and 48 and follow the procedure suggested above.

If recognizing perfect squares is a problem, return to Worksheet 27 again. Then try Worksheet 39 again, checking each problem before going on.

Post-Test. After you have completed this chapter, administer the post-test. Review the appropriate pages for any incorrect answers. Then retest. When S. has successfully completed the work, go on to the next book.

Introduction

If you have been referred to this book by the diagnostic test, have S. take the pre-test to determine where to begin in this book. On the answer page to the pre-test is a list of the questions and their related chapter pages. When S. has missed a question, turn to the chapter referred to and begin.

If S. is working through the complete series, there is no need to give the pre-test. The diagnostic test has already indicated a lack of skill, and it would be threatening and discouraging to have S. fail again at these questions. Simply begin with Chapter 1.

Chapter 1. Number Lines on Graphs

Concepts:

- Knowing about the *x*-axis and the *y*-axis
- Knowing about the origin
- Knowing how to name points on the *x*-axis, *y*-axis, and the origin

Pages 7 & 8. Point out to S. that the coordinates of the origin are (0, 0). Encourage S. every time she works with a graph, to label an unlabeled axis *x* to the right and *y* at the top. Read the discussion slowly with S., pointing and counting for the points mentioned. Each space between lines represents one unit. The lines mark the end of movement across units. The most common mistake made is counting the origin as "one."

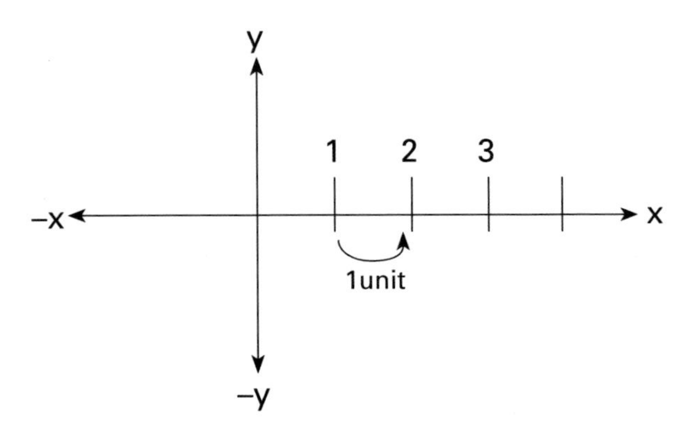

When showing S. how to count, put your pencil on the origin and mention that we must move off the origin to begin to count.

Example: "Move One." Draw a sample grid on the board. Show S. how to start at the origin and move one unit.

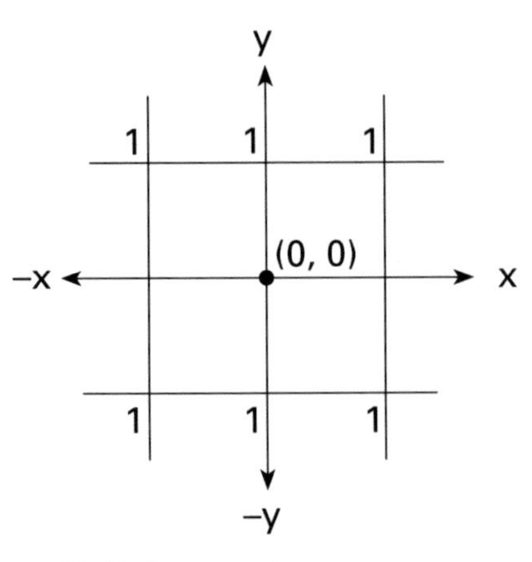

If S. is counting wrong, a good way to illustrate this is to have S. stand with her back against a wall. Then tell her to take two steps. Each step is the distance between where her left and right feet are. Point out that we begin to count as we put our foot down at the end of the first step. We don't count the origin because we don't count the place where we started as a step. The (0, 0) coordinates of the origin indicate that we have not yet moved either horizontally or vertically.

Page 9. Read the question and directions for Exercise 1 with S. Have S. do problem 1 and check it. If it is correct, have S. do the other problems, answering and checking each against the answer key before going on. If S. is still not completely successful, first look together at S.'s work to find the exact cause of difficulty. Then work together on the area of difficulty. Provide graph paper for S. to practice at home.

Activity

Materials: Worksheet 40 and pencil

Procedure:

Work this page with S. If she has any difficulty, go back to the discussion in the *Teacher's Guide* for pages 7 and 8 and reteach the method. Then try Worksheet 40 again.

Chapter 2. Finding Coordinates for Points

Concepts:

- Understanding coordinates
- Knowing how to use parentheses
- Knowing that the *x* comes before the *y* in the parentheses
- Knowing how to locate a point
- Knowing how to name the coordinates of a point

Pages 10 & 11. Have S. label the graph on page 11 as she did on Worksheet 40, using the (x, y) coordinates. Then read the discussion slowly with S., stressing the chapter concepts and pointing when appropriate.

Page 12. Read the directions to Exercise 2 with S. Have S. do problem 1 and check it. If it is correct, have S. do the other problems, answering and checking each against the answer key before going on. If S. is still not completely successful, first look together at S.'s work to find the exact cause of difficulty. Determine what steps are not correct. Then work together on the area of difficulty. If the counting method is the problem, return to the discussion in this *Teacher Guide* for pages 7 and 8 and reteach the method. Then try Exercise 2 again. Have S. copy on graph paper the problems and answers to practice at home (provide graph paper).

Activity

Materials: Worksheet 41, graph paper, and pencil

Procedure:

Work this page with S. If the counting method is the problem, return to the discussion in this *Teacher's Guide* for pages 7 and 8 and reteach the method. If S. needs practice reading coordinates correctly, go to the example in this *Teacher's Guide* for page 16 for the method of reading coordinates. Then try Worksheet 41 again. Have S. practice at home.

Chapter 3. Word Problems With Graphs

Concepts:

- Understanding how to translate word problems into equations
- Knowing how to read and plot points on a graph
- Recognizing math in everyday life

Pages 13 & 14. Read the discussion and ask for any questions or concerns that S. has. Do the example with S. and discuss how a graph is similar to a map.

Page 15. Have S. do Exercise 3. Have her plot Karem's path on the street grid in the book.

Chapter 4. Plotting Points

Concepts:

- Knowing how to plot the point given the coordinates using lines that intersect
- Knowing how to plot the point given the coordinates without drawing the lines that intersect

Pages 16 & 17. Read the discussion slowly with S., stressing the chapter concepts and pointing and counting when appropriate. Stress the last line on page 16. This style of reading coordinates seems to be most successful. On page 16, teach this style as follows:

Example 1: Read $(-5, -3)$ as "left 5, down 3."

Example 2: Read $(-3, 4)$ as "left 3, up 4."

Example 3: Read $(0, -2)$ as "no left or right, down 2."

Example 4: Read $(5, 0)$ as "right 5, no up or down."

Repeat this until S is completely successful and finds it easy. If S. has difficulty, try this additional explanation.

T: The coordinates are actually a set of directions for finding a unique point in space. The (x, y) is similar to telling a friend to meet you at the intersection of Main Street and West Street. In plotting points, remember to follow alphabetical order, counting first x, then y.

Page 18. Read the directions to Exercise 4 with S. Have S. do problem 1 and check it. If it is correct, have S. do the other problems, answering and checking each against the answer key before going on. If S. is still not completely successful, first look together at S.'s work to find the exact cause of difficulty. Determine what steps are not correct. Then work together on the area of difficulty. If the counting method is the problem, return to the discussion in this *Teacher's Guide* for pages 7 and 8 and reteach the method. If S. needs practice reading coordinates correctly, return to the examples in this *Teacher's Guide* for page 16 for the method of reading coordinates. Then try the exercise again. Have S. copy the problems and answers to do for practice at home (provide graph paper).

Activity

Materials: Worksheet 42, graph paper, and pencil

Procedure:

Work this page with S. If the counting method is the problem, return to the discussion in this *Teacher's Guide* for pages 7 and 8 and reteach the method. If S. needs practice reading coordinates correctly, return to the examples in this *Teacher's Guide* for page 16 for the method of reading coordinates. Then try Worksheet 42 again. Have S. practice this at home.

Chapter 5. Multiple-Choice Questions

Concepts:

- Understanding how to answer multiple-choice questions
- Understanding that wrong answer choices will be close to the correct answer

Explain that the wrong answers are often developed by following the correct answer steps but applying the common errors that people often commit. So, S. needs to be careful in doing computations and when reading the answer choices.

Pages 19 & 20. Read the discussion. Ask for any questions or concerns that S. may have. Draw the graph on the board and talk about positive and negative signs and where they occur. Go over the example, plotting each point after having S. decide where it should lie on the graph.

Page 21. Read the instructions to Exercise 5. Have S. answer all questions by filling in the bubble next to the answer.

Chapter 6. Plotting Equations

Concepts:

- Knowing that an equation may have many solutions with y changing as x is changed
- Recognizing and using a "T" table

x	$y = 3x + 2$	y	Points
1	$y = 3(1) + 2$	5	(1, 5)
0	$y = 3(0) + 2$	2	(0, 2)
–1	$y = 3(-1) + 2$	–1	(–1, –1)

- Understanding that any x may be used in the T table
- Using three points to ensure a correct straight line (or to help you see that you've made a mistake)

Pages 22–24. Read the discussion slowly with S., stressing the chapter concepts and pointing and counting when appropriate.

T: Many values for x could have been chosen to "plug in," but some other x's would not come out with neat whole numbers. So we try to pick those that do. A line is an infinite collection of points along a straight path. Since there is an infinite number of points and since they can't all be found, we use a few of them to enable us to draw the line showing all of the points.

As you complete each example, have S. try working it on graph paper to be sure the skill has been learned before you continue. If it has not been learned, discuss and have S. try it again. If the counting method is the problem, return to the discussion in this *Teacher's Guide* for pages 7 and 8 and reteach the method. If signs are a problem, return to Level 3 Book 1, Table of Contents to locate the type of sign-combining that is a problem. Review the necessary skills with S. before going on. If evaluating the equations for the chart solutions is a problem, return to Level 3 Book 2, pages 12 and 13. Review those pages, and then try these examples again.

Page 25. Read the directions to Exercise 6 with S. Remind S. to set up the equation three times, each time using each of the values given for x. Have S. do problem 1 and check it. If it is correct, have S. do the other problems, answering and checking each against the answer key before going on. If S. is still not completely successful, first look together at S.'s work to find the exact cause of difficulty. Determine what steps are not correct. Then work together on the area of difficulty. Have S. copy the problems and answers to do for practices at home. If solving equations is a problem, return to Level 3 Book 2, pages 12 and 13 and review the skills with S. before going on.

Activity

Materials: Worksheet 43, graph paper, ruler, and pencil

Procedure:

Work this page with S. If the counting method is the problem, return to the discussion in this *Teacher's Guide* for pages 7 and 8 and reteach the method. If signs are a problem, return to Level 3 Book 1, Table of Contents to locate the type of sign work that is a problem. Review the necessary skills with S. before going on.

Chapter 7. Solving Two Equations by Graphing

Concepts:

- Understanding that if two lines intersect, they have a common point called the *solution* (and are not parallel)
- Understanding that parallel lines never cross and therefore have no common point
- Knowing how to solve and plot two lines on the same graph

Pages 26–29. Read the discussion slowly with S., stressing the chapter concepts. Point and count, when appropriate, and find the equation solutions by using the T table. As you complete each example, have S try working it on graph paper to be sure the skill has been learned before you continue. If it has not been learned, discuss and have S. try it again. If solving equations is a problem, return to Level 3 Book 2, pages 12 and 13 and review the skill with S. before going on.

Page 30. Read the directions to Exercise 7 with S. Have S. do problem 2 and check it. Remind S. that working with the multiplicative inverse can help here:

$$4x = 2y$$

$$\frac{4x}{2} = \frac{2y}{2} \quad \text{(multiplicative inverse)}$$

$$2x = y$$

If problem 2 is correct, have S. do the other problems, answering and checking each against the answer key before going on. If S. is still not completely successful, first look together at S.'s work to find the exact cause of difficulty. Determine what steps are not correct. Then work together on the area of difficulty. If solving equations is a problem, return to Level 3 Book 2, pages 12 and 13 and review the skills with S. before going on. Have S. copy the problems and answers to do for practice at home (provide graph paper).

Activity

Materials: Worksheet 44, several sheets of graph paper, ruler, and pencil

Procedure:

Work these pages with S. If S. needs practice reading coordinates correctly, return to the examples in this *Teacher's Guide* for page 16 for the method of reading coordinates. Then try the exercise again. In problem 2, remind S. that working with the additive inverse can help.

$$\begin{array}{rcl} x + y = & & 3 \\ -x & & -x \\ \hline y = -x + 3 & & \end{array} \quad \text{(additive inverse)}$$

$$\begin{array}{rcl} 2x + y = & & 5 \\ -2x & & -2x \\ \hline y = -2x + 5 & & \end{array} \quad \text{(additive inverse)}$$

Have S. practice this at home. If the counting method is the problem, return to the discussion in this *Teacher's Guide* for pages 7 and 8 and reteach the method. If signs are a problem, return to Level 3 Book 1, Table of Contents to locate the type of sign work that is a problem. Review the necessary skills with S. If evaluating the equations for the chart solutions is a problem, return to Level 3 Book 2, pages 12 and 13. Then try Worksheet 44 again.

Post-Test. After you have completed this chapter, administer the post-test. Review the appropriate pages for any incorrect answers. Then retest. When S. has successfully completed the work, proceed to Level 4.

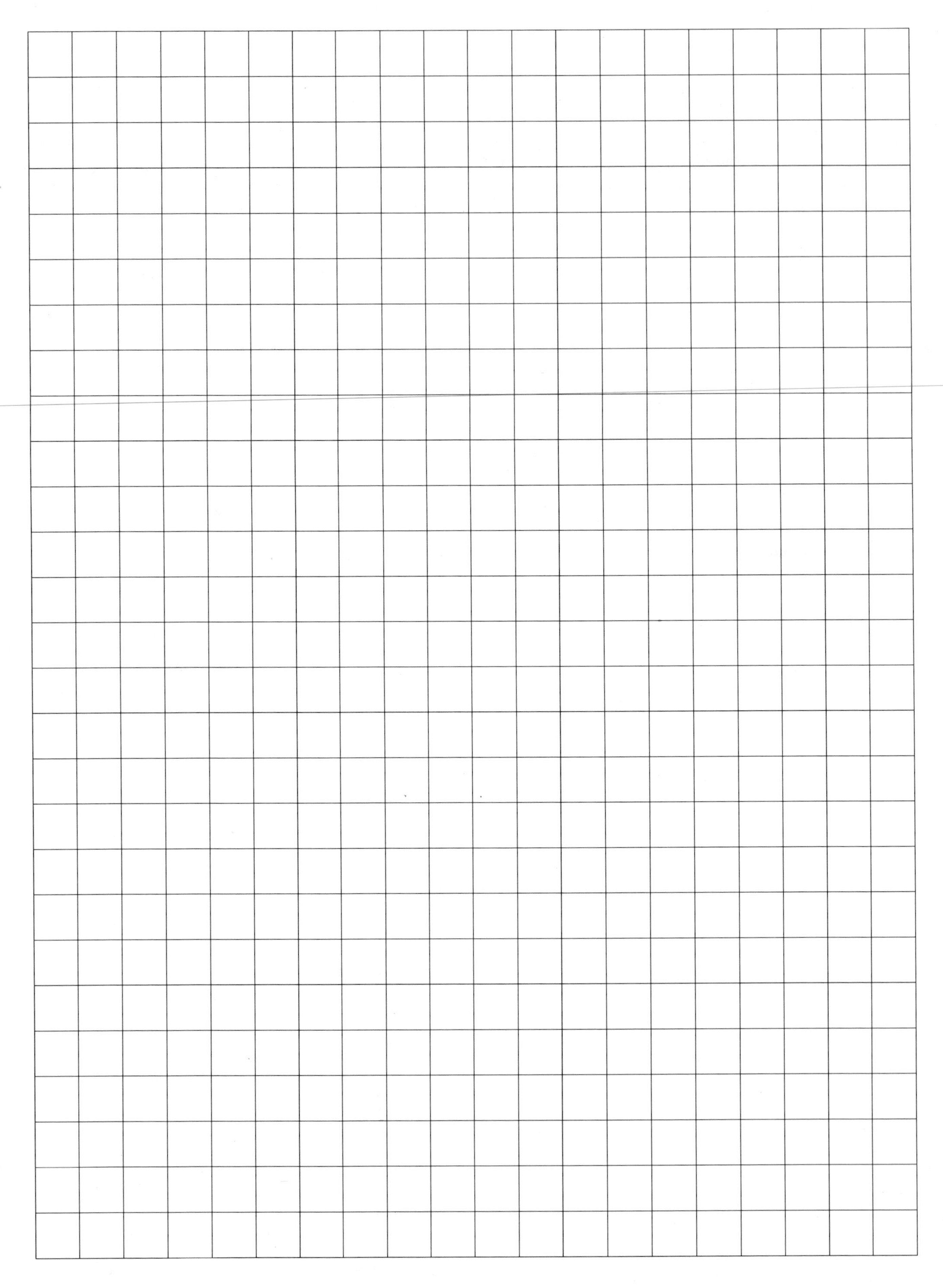